T0210645

Medicine as a Scholarly Field: An Introduction

O.S. Miettinen

Medicine as a Scholarly Field:
An Introduction

 Springer

O.S. Miettinen

Department of Epidemiology, Biostatistics
   and Occupational Health, and Department of Medicine
Faculty of Medicine
McGill University
Montreal, QC, Canada

Department of Medicine
Weill Medical College
Cornell University
New York, NY, USA

ISBN 978-3-319-36864-1   ISBN 978-3-319-19012-9   (eBook)
DOI 10.1007/978-3-319-19012-9

Springer Cham Heidelberg New York Dordrecht London
© Springer International Publishing Switzerland 2015
Softcover reprint of the hardcover 1st edition 2015

Printed on acid-free paper

Springer International Publishing AG Switzerland is part of Springer Science+Business Media (www.springer.com)

Foreword

Careful development, definition, and use of concepts, while firmly ingrained in the discipline of philosophy, should also be an integral feature of any scholarly field, medicine included. The deployment of well-developed concepts is essential for critical thinking and, hence, of central importance to scholarly progress. Medicine, like other scholarly fields, stands to benefit from improved concepts and principles and from better-developed and more explicit philosophy; and this is precisely what is provided in this important new book by Olli Miettinen.

Miettinen's point of departure is his perception of an important but generally missing module in medical education; and this book provides that missing element: an introduction to medicine at large, to its concepts and principles, and also to the philosophy beyond these essentials, relevant to all specialties of medicine and written with precision, lucidity, and insight. Definitions of core concepts of medicine are proposed; terms which often are unreflectively used interchangeably are distinguished from one another; and propositions are put forward concerning the central logical and ethical principles of medicine. The book moreover presents principles for the pursuit of professional excellence and professional happiness of those who have, or will have, careers as physicians.

Taken in its entirety, Miettinen's new book provides an important philosophical foundation for medicine. As such, it is not a book to be read quickly, but to study, to contemplate, and to return to. The reader who wrestles through the pages of this dense text will be rewarded by Miettinen's stimulating prose, probing style, and call for action and, most importantly, will emerge with a clearer, sharper understanding of the core concepts and principles of medicine.

This book offers compelling insights to anyone who reads through its pages and follows its arguments; there is something worthy of consideration for everyone within this introduction to medicine as a scholarly field. The book unveils the logical structure of the rational practice of medicine and the nature and sources of the requisite knowledge-base for it, however lacking that knowledge may still be. It clarifies the relationship between medical research, in all its forms, and the arts of the practice of medicine. It posits core principles for the ethics of medicine. And it outlines the historical evolution and present state of medicine and advances visions of future improvements in the field. This book is thus essential reading for clinicians, clinical researchers, and epidemiologists and even for medicine's basic scientists. For physicians-to-be, as medical students, it provides what it itself argues has been lacking, an introduction to the concepts and principles of, and philosophy behind, medicine as a scholarly field, and furthermore provides guidance on how relevant principles can serve the development of their future careers in medicine. And for practicing physicians this book offers an opportunity to step back, to reconsider general tenets of medicine sometimes held as immutable, to critically reflect upon various experiences in one's own practice, and perhaps even to help alter the course of medicine itself. The book offers an alternative vision of what medical practice should be.

Throughout, Miettinen beautifully elucidates the concepts and principles of knowledge-based diagnosis, and prognosis, within medicine. Now, after six decades of keen observation and study, and critical reflection on medicine and medical research, Miettinen, in this book, shares the fundamental understandings he has reached and provides his own diagnosis of medical practice itself. The prognosis of physicians' professional excellence, and professional happiness, may well depend on how thoroughly his precepts are learned and his prescriptions heeded.

Harvard University, Cambridge, MA, USA   Tyler J. VanderWeele

Preface

Any modern student of medicine, at the outset of the studies already, knows and understands a fundamental truth about modern medicine as a professional field: Consequent to its ever greater complexity, medicine has fragmented into many constituent disciplines – 'specialties' and 'subspecialties' – of it; and so, any given doctor now practices only in some limited segment of medicine, 'general practice' (a.k.a. family medicine) being a misnomer for one of the limited disciplines of modern medicine. The student thus is not setting out to learn medicine at large but only some particular discipline within it.

The student also knows, at the outset already, something equally notable, though much less comprehensible, about the educational preparation for whichever one of the disciplines of modern medicine: For access to studies specific to the particular discipline of his/her future practice of medicine – be it dermatology or diabetology or psychiatry or whatever else (with dentistry commonly the exception) – (s)he is required to complete an education leading to the MD (Medical Doctor) degree (in the USA and Canada) or some equivalent of this (elsewhere).

This requirement implies that such a degree signifies mastery of what I call *the medical common* – the educational content that is relevant to each of the constituent disciplines of modern medicine – all of this and nothing but this, as best, is understood by the authorities who specify the required contents of medical-school studies in a given jurisdiction of medicine at a given time. For the USA and Canada, the required studies are now stipulated by these countries' Liaison Committee on Medical Education. The appointment of its members and its current composition are specified in lcme.org.

These undergraduate studies for all of medicine now commonly involve, inter alia, introductions specific to some of the currently official constituent disciplines of medicine. Whether these modules, or any of the others for that matter, actually merit inclusion in the medical common is not a concern of mine in this particular book. Instead, I here focus on what I regard as a much-needed but generally missing module in these studies, *introduction to medicine at large*, and this in such terms as are *relevant to whichever genuine discipline of medicine*, regardless of whether the discipline is officially recognized or even exists as some doctors' career focus in medicine. That is, the focus here is on introduction to medicine at large as a prospective module early in undergraduate medical education – wherever, whenever.

Such an introduction to medicine would address the general *philosophy* of the field, the modern counterpart of the philosophy – Hippocratic – at the root of the genesis of modern medicine. It thus would introduce, in its first phase, critically formed (ontologically 'real') general *concepts* of the objects of medicine and of medicine itself, naturally together with apposite terms denoting these (in the lingua franca of the modern world wherever English is the dominant language, but otherwise in the local counterpart of this or both). It would then introduce general *principles* of medicine, both logical and ethical. And as these principles would call for the deployment of medical *knowledge*, introduction to the nature (ontic) and sources (epistemic) of this knowledge would be included. Besides, the introduction might well outline the avenue to the attainment of true *professionalism* in modern medicine – as to how, to this end, a genuine discipline of medicine is best defined and how full competence in a well-defined discipline of modern medicine is best pursued.

The need for such a module in undergraduate medical education is evident from the prevailing confusion about those introductory matters of medicine, which this book exposes – and endeavors to rectify. The status quo implies, more specifically, that medical students need an introduction to medicine at large on a *sufficiently high level of scholarship*, for it to help forestall such confusion among doctors of the future.

To wit, I see a need for an introduction to medicine at large such that from it any student preparing for a career in whichever genuine discipline of medicine would learn to think, critically, about the answers to questions such as: Are the terms 'disease,' 'sickness,' and 'illness' (in English) synonyms? What about 'treatment,' 'therapy,' and 'intervention' in this respect? What is true and unique about all genuine disciplines of medicine, distinguishing them from all paramedical professions? Can a modality of treatment – surgery, say – be definitional to a genuine discipline of medicine? What logically is the essence of diagnosis, and what is the source of its requisite knowledge-base? What are tenable conceptions of scientific medicine and medical science, respectively? What is the essence of ethical medicine? How is professional happiness in medicine best assured?

Scholarly thinking about the answers to these questions and much more, any student of medicine would learn from sufficiently well-developed general philosophy of medicine addressed in *propaedeutic studia generalia* – introductory general studies – of the field. Whereas the prevailing confusion about these introductory matters of medicine reflects the still-continuing absence of such studies from medical education (and vice versa), it is by no means a priori clear even to the clinical members of the faculties of medical schools what the topics and especially the actual contents of these studies should be.

In this book I posit propositions not only on the broadest syllabus but also on the core contents of these studies – for leaders, providers, and, especially, recipients of medical education to 'Read not to contradict, nor to believe, but to weigh and consider' (ref. in Section 2.2). These propositions derive from, and reflect, my serious efforts, over almost six decades by now, to understand the big picture of both medicine and medical research – efforts motivated by my sense of the importance of these understandings for properly purpose-driven and genuinely insightful innovations not only in medical education but also in medicine proper and the research to advance this preeminent one of the 'learned professions.'

Only now, after so many decades in medical academia with that outlook, do I feel ready to assume this task, more challenging than any of the ones I have previously undertaken. But this does not mean that comprehension of the introduction to medicine in this book is beyond the reach of beginning students of the field, provided that they – by their suitable selection – have the right attitude: appreciation that determined, serious effort is needed to acquire the 'essential competencies' of medical professionals specified in the widely adopted CanMEDS document, one of these being *competence as a scholar* in one's particular discipline of modern medicine.

This exalted professional status – the genuine version of it – cannot be attained in the framework of what Frank Furedi laments as the now-prevalent 'twenty-first-century philistinism' in academia. Where this is the quality of the surrounding academic culture, students of medicine need not – and should not – assimilate it into their own outlook as students and, ultimately, as doctors.

Montreal, QC, Canada   O.S. Miettinen

Acknowledgments

The words in all the drafts, and in the final version too, of the manuscript of this book were 'processed,' competently and cheerfully, by *Chantal Burelle*. This contribution to the genesis of this piece of work was a sine-qua-non for it, given the continual agenesis of such skills in me.

Three colleagues at my current home university (McGill) read the first draft of this text: *Kenneth Flegel* (internal medicine and editorship of a medical journal), *Richard Cruess* (orthopedic surgery and deanship of a medical faculty), and *Sylvia Cruess* (internal medicine and endocrinology) – the latter two now eminent members of our medical faculty's Centre for Medical Education. The feedback from all three of these colleagues was very encouraging, thus constituting the other sine-qua-non for the genesis of this book.

That pivotal set of encouragements was supplemented by the sole reader of an early draft at the university of my adjunct affiliation: *James Smith* (pulmonology) at Cornell University. He also made valuable suggestions for improvement of the comprehensibility of the expositions of some of the more subtle clinical concepts, novel ones in particular.

Igor Karp (epidemiology) at Université de Montréal and University of Western Ontario read practically all drafts of this text and helped weed out various technical and some conceptual deficiencies. He also checked the statistical results presented in Section 8.1 and Appendix 2.

Alfredo Morabia (epidemiology and history of it) at City University of New York and Columbia University read a late version of the manuscript and made a valuable contribution to the focus of Chapter 1 (on the genesis of modern medicine) and, thereby, to the cohesiveness of the text as a whole. He described the resulting text at large as having the structure of a fugue (à la J. S. Bach).

By far the principal one among the colleagues contributing to the contents of this book has been *Johann Steurer* (internal medicine) at the University of Zurich. As the head of the Center for Patient-Oriented Research and Knowledge Transfer there, he has, for many years, been transferring relevant knowledge to me in the form of a large number and variety of books, thus greatly enriching many parts of this text.

Indeed, many of the books cited in this text I came to study on account of this highly learned and exceedingly generous colleague. He also read some drafts of this text and made valuable suggestions. And about this text overall he remarked: "To me it is astonishing and for medicine actually shameful that it has taken up to year 2015 before there is a work in which the essence of medicine is described and discussed."

A late draft of this text was read by someone with a very different but here centrally relevant type of expertise: *Ilkka Niiniluoto* (theoretical philosophy and rectorship of a university) at my original alma mater, University of Helsinki. The opening of his feedback was this: "The aim of this book ... is admirable. The composition of the work – from the key concepts to logical and ethical principles – is very clear and systematic. I am convinced that this kind of book is needed." On some of the topics he stimulated worthy additions. And one of his asides I found to be particularly notable: "The diagnostic method [*sic*] ... is still hotly debated by philosophers of science." So, confusion about some matters of the theory of medicine prevails among philosophers too, but they evidently are well aware of this and active in seeking to correct it.

A practically final version of the manuscript was studied by *Tyler VanderWeele* at Harvard University. My interest in his reading of that version of the manuscript – when inputs for improvements no longer were my concern – was based on his being a veritable scholar in the health fields with a here-relevant distinction in this: Like any student entering medical school, he has not undergone studies leading to the MD degree or any equivalent of this. I thus judged him to be particularly well suited to produce a Foreword for this book, to thereby orient beginning students of medicine to the contents here specifically from their vantage – to what students approaching medicine with the requisite scholarly bent (Preface) can expect to learn from this book and to what avail in the pursuit of competence as scholars in their particular future disciplines of modern medicine. As is evident, he agreed to carve out the time to render this special service to the cause here.

To say something that would go without saying, I am indebted to everyone herein acknowledged and to many others besides.

Contents

Part I
Preamble

Chapter 1
The Genesis of Modern Medicine

Contents

1.0 Abstract

The genesis of modern medicine is generally taken to have begun with the teachings of a master physician in ancient Greece, named Hippocrates. The empirico-rational core of these innovative teachings of medicine was rooted in Greek philosophy, but they also embodied, centrally, a purely speculative doctrine about the nature of human maladies.

The Hippocratic teachings were much extended, with that doctrine upheld, by Galen in Rome, half-a-millennium later. That doctrine misled medicine, seriously, till quite recently. But nevertheless, the Hippocratic philosophy of *practice-based learning* has been the basis for the development of quite substantial a knowledge-base for medicine.

The recent advent of *medical science* in particular has served to make a wealth of 'tools' to be available to physicians. This has greatly increased the complexity of the requisite knowledge-base of medicine – and this, in turn, has led to inescapable *fragmentation* of medicine into differentiated disciplines of it (cf. Preface).

© Springer International Publishing Switzerland 2015
O.S. Miettinen, *Medicine as a Scholarly Field: An Introduction*,
DOI 10.1007/978-3-319-19012-9_1

1.1 Hippocrates as 'The Father of Medicine'

The medicine of Greece in the fifth century BCE was, still, mostly under the spell of the healing god Asclepius, for whom there were hundreds of temples in the country. Sick people were selectively admitted to a healing ritual called incubation. While in sleep, they were visited by *Asclepius* himself, incarnated as a priest, together with his daughters *Hygeia* and *Panacea*. These healers effected their chosen treatments, and by the morning the (suitably selected) patients commonly were cured.

As an alternative to this 'temple medicine' there arose, in Greece at that time, '*empirical medicine.*' In this radically new type of medicine, inspired by Greek philosophy, maladies were no longer viewed as punishments visited upon people by gods. They became regarded as natural phenomena, and specifically as manifestations of imbalances/disharmonies among the *four 'humors'* of the body – blood, phlegm, cholera or yellow bile, and melancholy or black bile – and healing too got to be seen as the result of natural influences, ones that restore balance among these 'humors.' Very importantly, these phenomena and the influences on them were taken to be subject to understanding by physicians, through their *observations and rational thinking.*

This philosophical revolution in medicine has been attributed to a master physician named *Hippocrates* (ca. 470–ca. 380 BCE). Very little is now known about his life; and he – like Socrates – may never have written anything. But there evolved a large body of 'Hippocratic' writings – the Hippocratic Collection or *Corpus Hippocraticum* – which became the library of a medical school and was later transferred to the famous library in Alexandria, where it was edited further and made generally available to scholars. Best known of these writings now are the Hippocratic Oath (Section 6.4) and one of the Aphorisms: "Life is short, the art long; timing is exact, experience treacherous, judgement difficult." (The various translations vary somewhat.)

The temple medicine of ancient Greece is the basis of the *modern symbol of medicine*. In its adoption, the intent was that the symbol represent the rod of Asclepius, around which is wound one serpent; but by a misunderstanding, the symbol got to be the rod of Hermes – the messenger god, with medicine in his domain – which has two serpents, not only one. However, even with the right number of serpents, the rod of Asclepius would be a false symbol of modern medicine. For *the essence of modern medicine* is decidedly non-Asclepian; it is Hippocratic in philosophy, committed to empiricism-cum-rationality with no connection to mythology.

In his magnificent treatise on the history of medicine (ref. below), Felix Marti-Ibañez characterized the Hippocratic physician as a "learned" one, "a man wise, modest and humane" (p. 66). A modern physician, when excellent, is Hippocratic in that wholesome meaning Marti-Ibañez associated with this adjective, meaning that harks back to '*The Father of Medicine.*'

Reference: Marti-Ibañez, F. *The Epic of Medicine*. New York: Clarkson N. Potter, Inc., 1962.

1.2   Galen as 'The Second Hippocrates'

A leader of medicine equal in statue to Hippocrates emerged half-a-millennium after him. He was *Claudius Galenus* (ca. 130–ca. 200 CE), or simply Galen. Highly educated in medicine and philosophy in the Hellenic world (and arrogant and ambitious to boot), he moved to Rome and made his career – highly successful – there. *He upheld the Hippocratic doctrine of the four 'humors,'* of their central role in human health; and *his medical writings – very extensive – became canonical and remained so throughout the Middle Ages, for 14 or 15 centuries.* He thus could be dubbed 'The Second Hippocrates.'

Galen's teachings were hotly debated in the seventeenth century, but this did not bring an end to the preeminent mode of treatment of diseases in Galenic medicine: *bloodletting,* to regain the balance among the four 'humors.' At that time, still, as is noted in a recent book on the history of medicine (ref. 1 below), *Thomas Sydenham* (1624–1689) – 'The British Hippocrates' – "began the treatment of virtually every disease by opening a vein with a lancet" (p. 16). And even much later, this: On the day George Washington died (in 1799), he was treated (by eminent doctors) with "repeated bleedings – over half of [his] circulating blood" (p. 76).

Reference 1: Abrams JE. *Revolutionary Medicine. Founding Fathers and Mothers in Sickness and in Health.* New York: New York University Press, 2013.

The 'humoral' conception of the essence of maladies, à la Hippocrates and Galen, was given an eminent alternative by *Rudolf Virchow* (1821–1902), in Berlin. Instead of imbalance among the 'humors' in the body, he viewed maladies as representing cellular reactions to noxious stimuli, as set forth in his epoch-making *Cellular Pathology* (1858). But as for the practice of medicine, an illness quite similar to that which was fatal to George Washington (acute, respiratory) was terminal and treated with bloodletting ('cupping') also in the case of Ivan Pavlov – the physician-physiologist and Nobel laureate – in 1936 (ref. 2 below).

Reference 2: Todes DP. *Ivan Pavlov. A Russian Life in Science.* New York: Oxford University Press, 2014. P. 723.

1.3   Hippocratic Progress in Medicine

Given the Hippocratic commitment to rational empiricism in the practice of medicine, the question now is: What progress in the practice of medicine has this philosophy actually brought about? That is: How has the *knowledge-base* of medicine grown on account of practitioners' keen observations and rational thinking about these? This question arises in particular because physicians' thinking about the enormous experience with bloodletting (by 'venesection,' 'cupping,' or 'leeching'), founded on the Hippocratic-Galenic doctrine of 'humors,' did not result in recognition of its uselessness and, even, counterproductivity (Section 1.2).

Despite that outstanding failure, in which an a-prioristic doctrine trumped learning from experience, there are innumerable examples of the Hippocratic-type, practitioner-driven progress in medicine. Very important was the thus-gained collective understanding of some diseases as being communicable from person to person, which translated into effective preventive measures; and acquired immunity became recognized and made use of, most notably in 'inoculation' against smallpox. This practice was adopted early on in the Orient, outside the reach of Hippocratic teachings, while only relatively late and reluctantly in Britain and America (ref. 1 in Section 1.2).

Related to this is the greatest one of a single practitioner's contribution to the knowledge-base of medicine: *Edward Jenner* (1749–1823), a country practitioner in Britain, made the shrewd observation that persons who had experienced cow pox – a very mild disease – did not contract smallpox – a very common and commonly fatal communicable disease. He postulated that deliberate transmission of matter from a cow-pox pustule to a person who has not yet experienced smallpox (nor cow pox) might prevent this person from subsequently contracting smallpox (just as a non-fatal case of smallpox itself was known to do). He experimented on this, on a single subject (in 1796), and this persuaded him of the correctness of the idea. Others were persuaded, quite grudgingly, by his further work. In America, an early champion of the vaccination based on this discovery was Thomas Jefferson (ref. 1 in Section 1.2). By now, smallpox has been completely eradicated from the whole world, vaccinated out of existence. (The etymology of 'vaccine' has to do with cow.)

Another outstanding discovery by an individual practitioner was the 'iatrogenic' (physician-induced) nature of the very high mortality from 'puerperal' (childbed) fever in hospitals, which became preventable by mere hand-washing by hospital physicians – commonly returning directly from the autopsy room to the maternity ward – before conducting pelvic examinations of the expectant mothers. This contribution to the knowledge-base of medicine was made, by inference from rates of occurrence of this malady in hospital practices, by *Ignaz P. Semmelweis* (1818–1865) in Vienna, and independently of this by *Oliver W. Holmes* (1809–1894) in Boston.

The medical establishment's resistance to accepting this very important, firmly evidence-based idea was very great, including by Rudolf Virchow (Section 1.2) – greater even than the resistance to the epoch-making idea of Jenner's.

1.4 Scientific Progress in Medicine

1.4.1 'Basic' vs. 'Clinical' Contributions

Hippocrates could not have had any ideas about science as a driver of progress in medicine; for, apart from astronomy, empirical science had not yet been introduced (by Aristotle), to say nothing about specifically medical science. (The theoretical science of mathematics did already exist.)

Science is either 'pure' or 'applied,' meaning that it seeks knowledge either for the sake of knowledge itself or for its application to practical ends. *Science that properly is characterized as medical is applied science*; it is intended to advance the arts of medicine, physicians' provision of healthcare.

Medical science of a given kind advances the arts it serves in one of two fundamental ways: it either helps to bring, through the research (followed by appropriate 'development'), novel *'tools'* for possible use in medicine (diagnostic tests and therapeutic medications, most notably), or it improves the *knowledge-base* of medicine (for the use of those 'tools,' i.a.). The corresponding two broad lines of medical research are commonly labelled (rather questionably; Section 1.4.3) as *'basic'* and *'clinical,'* respectively. Research for the knowledge-base of practice I think of as the most concretely applied – quintessentially applied – segment of medical research.

'Basic' – laboratory-based – medical research has already led to an enormous number of new products and processes for consideration by doctors in their practices. On the other hand, alas, 'clinical' research – in the human domain – has remained quite unproductive in the advancement of the knowledge-base of medicine, notably that of setting suitably individualized diagnostic and prognostic probabilities – the latter, most importantly in modern medicine, also as to their dependence on the choice of treatment. The scientific knowledge-base for such case-specific, individualized probability-setting remains, alas, practically non-existent, as the requisite research has scarcely begun.

1.4.2   Academic vs. Industrial Contributions

It is commonplace to hold that prior to the programmatic, collaborative, largely industrial medical research that emerged in the eighteenth century, major scientific contributions to medicine had already been made by certain individuals in academia, solo, most notably by *Andreas Vesalius* in the sixteenth century and *William Harvey* in the seventeenth. But their contributions were to anatomy and physiology, respectively, not to medicine. (Knowledge about normal bodily structures and functions, however complete, does not constitute the requisite knowledge-base for medicine: a doctor practicing medicine need not know about matters such as the Watson and Crick double helix structure in the genome, or the Krebs cycle function in intermediary metabolism – both of which have been the basis for Nobel prize in "Physiology and Medicine.")

Medical research (defined in Section 1.4.1) as an organized endeavor can be said to have begun with an innovation in industry: the formation of a pharmaceuticals-oriented offshoot of the chemical company IG Farben in Germany. The founder of this novel type of company was *Friedrich Bayer* (in collaboration with Johann Friedrich Wescott). At its founding, in 1863, it was named Bayer & Company, and now it is the gigantic Bayer Group.

Even today, 'basic' medical research is principally directed to the development of *pharmaceuticals*, and it still is principally conducted by industry rather than academia. (Much of the purportedly medical research in academia is biological pure research, rather than medical research, inherently applied; cf. Section 1.4.1).

Very regrettably, research to develop and advance the *knowledge-base* of clinical practice is failing to thrive, as industry lacks the profit motive for it and academia is confused about its theoretical framework (Appendix 1).

1.4.3   Epidemiological Contributions

In medicine at large, a major duality is based on whether the doctor serves individuals, one at a time, or the population of a community, as a population (rather than its individuals separately). This is the duality constituted by clinical medicine versus *community medicine*; that is, clinical medicine versus *epidemiology*.

Scientific contributions to progress in these two broad branches of medicine (Section 4.1.3) have been made by what reasonably are called clinical research and *epidemiological research*, respectively. The clinical research advancing clinical medicine has been and is laboratory-based on one level, and clinically-based on the more directly applied level (Section 1.4.1). A corresponding duality obtains in epidemiological research as well: epidemiological research is either *laboratory-based* or *population-based*.

The scientific contributions to the advancement of medicine have, up to now, been more substantial on the population/epidemiological level than on the clinical level; and even the epidemiological advancements have derived more from laboratory-level research than from its population-level counterpart. The greatest progress in medicine has been the achievement of population-level prevention of various *communicable diseases* – by means of *vaccines* developed in the light of the results of laboratory-level epidemiological research.

Population-level epidemiological research on communicable diseases began before any of the microbial agents of these diseases had been discovered, with the discovery of the water-borne spread of *cholera*. The basis for this discovery was the 'observational' (non-experimental) work of the British doctor John Snow. This discovery became the inspiration for extensive 'epidemiological' research for the identification causes of illnesses with a view to the development of preventive measures. Particularly notable have been the identification of various *occupational exposures* as causes of illness and of *smoking* as a cause of lung cancer in particular.

1.5   Progress Leading to Fragmentation

In Sections 1.3 and 1.4, the mantra was 'progress in medicine,' and the big picture was sketched as one of great progress from antiquity to modernity. This progress was attributed to two principal factors: rational thinking about experience in practice (in the framework of the Hippocratic philosophy of progress in medicine) and medical science (cum technological development). From the latter source, as I noted, a great wealth of 'tools' has become available for (selective) use in the practice of medicine, diagnostic and therapeutic ones in the main.

A consequence of the progress – the scientific progress in particular – was a novel challenge: the *need for an ever-more complex knowledge-base* for medicine, for setting (tests-dependent) diagnostic and (treatments-dependent) prognostic probabilities.

In the face of this ever-mounting challenge, a modern doctor imbued with the Hippocratic virtue of wisdom (Section 1.1) adopts two post-Hippocratic principles. The first of these is the *need to focus on medicine*, on the knowledge (and skills) the possession of which constitutes competence as a doctor, thus leaving medical science to medical scientists to focus on. The wise doctor understands that familiarity with the science at the root of the development of any one of the 'tools' or, even, with the research behind the practice-relevant knowledge about the 'tool' itself, is not relevant for a practitioner to know about, much less to understand. The pursuit of familiarity with and understanding of this science (s)he understands to be a distraction from the pursuit of mastery of the actual knowledge-base of practice (of setting diagnostic and prognostic probabilities; Section 1.4).

The second, supplementary principle naturally is the *need to focus on a suitably narrow segment of medicine* (cf. Preface) and, ideally, one whose requisite knowledge-base is suitably coherent – so that knowledge relevant to a given subsegment of the practice is also relevant to the others (at least some of them). As it is, no one takes the entirety of the therapy-oriented – therapeutic – segment of medicine to be their domain of expertise; but incongruously with this, some have their expertise in modern medicine officially defined in reference to its preventive segment without any restrictions in this!

Overall, thus, a wise doctor practicing modern medicine pursues *excellence* of his/her practice in the framework of sharply distinguishing between medicine proper and medical science, focusing on medicine proper; and within medicine, (s)he focuses on a segment of it that is narrow enough and also otherwise such that (s)he can achieve an unsurpassed level of competence in it. Such a focus (s)he understands to be an important means to the end of achieving and maintaining excellence in his/her professional services – and, thereby, happiness as a professional.

Chapter 2
How to Introduce Modern Medicine

Contents

2.0 Abstract

In this book I address introduction to medicine in the spirit of taking it to be a needed module in the curriculum of a medical school, for common study before the rest of the education, and training too, relevant for a career in whichever particular discipline of modern medicine (cf. Preface). There is no preexisting model for this.

I take it that the introduction of future doctors to their professional teachings in their upcoming practices of medicine should now be as erudite as would be an exemplary introduction of future rabbis to their counterpart of this, and also in accord with the medical implications of the teachings in the philosophical disciplines of logic and ethics. I thus take it that the introduction should have *a tenable concept of ill-health and also of medicine itself as its two points of departure*; and that from these core concepts it should deduce, logically, various other, derivative concepts. And closely related to this, I take it that the introduction should address, also, correct use of these concepts in medical thinking, that is, the general *principles of medicine*, logical and ethical.

Besides, in an introduction to modern medicine it would be good, I think, to orient the student to the attainment of true *professionalism* in whichever genuine discipline of modern medicine.

© Springer International Publishing Switzerland 2015

O.S. Miettinen, *Medicine as a Scholarly Field: An Introduction*,

DOI 10.1007/978-3-319-19012-9_2

2.1   A Paradigm for the Introduction

Even though there already is an enormous literature on 'medicine,' there still is no unambiguous and shared conception of what medicine is. 'Medical' dictionaries define medicine in reference to physicians' practice of healthcare but are mutually discordant and also otherwise amiss about the specifics in this (Section 4.1.1); textbooks of 'medicine' address particular illnesses ('diseases') much more than physicians' practice of healthcare; and journals of 'medicine' – and 'medical' journals – imply by their contents that medicine is, more than anything else, a species of science, of scientific inquiry. Unsurprisingly, thus, there is no existing model, within medical academia itself, for how to introduce medicine at large to students preparing for careers in particular ones of its constituent disciplines: there is no such introduction.

There is an analogy between a *doctor* attending to the health concerns of people and a *rabbi* attending to their spiritual concerns. For both 'doctor' and 'rabbi,' the word's etymology has to do with the professional's essence as a *teacher*; and this is fitting because in both modern, 'scientific' medicine and modern, Talmudic Judaism the content of the teaching commonly is quite esoteric, inaccessible and/or incomprehensible to a lay person without the personal teaching of a learned professional. So, an exemplary introduction of rabbis-to-be to the eminently scholarly field of Judaism would, in certain respects, serve as a paradigm for an appropriate introduction of doctors-to-be to medicine as a field more scholarly than it is at present (Preface).

An exemplary introduction to Judaism is implicit in a medieval text addressed in a recent book (ref. below) about that text's author: Moses Maimonides (1135–1204), the physician and philosopher who, from this background, developed into his recognized status as the foremost intellectual in medieval Judaism, as 'The Second Moses.'

Reference: Halbertal M. *Maimonides. Life and Thought*. Princeton: Princeton University Press, 2014.

The text by Maimonides I am referring to is entitled *Misneh Torah*, which Halbertal translates as "Repetition of the Law" (p. 11). The Torah – this Hebrew word, as Halbertal (i.a.) says, means 'instructions' or 'teachings' (cf. 'rabbi' above) – is otherwise known as the Pentateuch, and it is the first of the three parts of the Jewish Bible (the *Tanakh*) – the latter, in turn, constituting the Old Testament in the Christian Bible (with the order of the second and third parts changed). And that text by Maimonides is a repetition of the teachings in the Torah – Mosaic Law, that is, rather than Jewish law at large – in a particular scholarly meaning of 'repetition.' For example, two separate prohibitions in the Torah Maimonides viewed as particular implications of a single, higher law he himself formulated, which stipulates the landowner's "obligation to leave produce in the field for the poor" (p. 109).

Maimonides viewed the Mosaic teachings in the Torah, which he thus "organized," as *canonical*, as he saw them to be generally agreed upon as the basis of the entirety of Jewish teachings, Jewish law. Following those canonical teachings – generally viewed as divine revelations – comes the human *interpretation* of those received laws, leading to new normative content in Judaism (p. 106). In this rabbinical extension of the biblical laws, Maimonides emphasized the need to assure "consistency of the new norm with the norms given earlier, at Sinai" – in the "deduction" of the new from the old (p. 106). This consistency he did not see to be the case with much of the rabbinical commandments in the then Talmud (p. 117).

Maimonides, a scholar akin to a (Constitution-interpreting) Supreme Court justice as he was, distinguished between two types of interpretation of the canonical. One of these he took to be a matter of seeking "to clarify or disclose the meaning of the text's terms," and it was to be guided by express "*principles of discovery and definition.*" The other "does not explain the text, but draws additional conclusions from it." (Pp. 120–121.) This latter type of interpretation was to be guided by express "*principles of deduction*" (p. 123). But the deductions had remained subject to disputes (p. 126).

Maimonides also wrote about *Jewish life* in the context of Jewish laws:

[The laws are] the means that enable a man to fulfill his purpose as a knowing creature. The Torah commands not only the construction of a proper society and the development of moral attributes, which are the means of attaining human perfection. Beyond that, it commands proper beliefs and involvement in contemplation as a way toward man's higher perfection. [P. 139.]

In the education of a rabbi-to-be, an *introduction to that which a rabbi teaches –* to Jewish law and Jewish life, that is – could, thus, focus on the philosophical *theory* – concepts and principles – of interpretation of the canonical, Mosaic Law, as this is tantamount to introduction to the Talmud ('study,' 'learning') at large, specifically to the reformulation/'summary' of this in the form of the *Misneh Torah*. But so as to be suitably practical about this, introduced could also be the *role of the law* at large in Jewish life, including in a rabbi's personal life.

2.2 Implications of the Paradigm

The introduction of a doctor-to-be to medicine – to his/her work as a teacher of people about their health – could already be rather analogous to an exemplary introduction of a rabbi-to-be to Judaism (Section 2.1 above) – to his work as a teacher of people about their living the Jewish life – if medical scholarship had already reached the level of its Talmudic counterpart, especially the Maimonides' version of this.

If medical scholarship were suitably advanced, there would already be, for one, a generally respected, if not canonical, body of writings on the *concept of health* (inclusive of ill-health) and on its subordinate concepts (those of illness and sickness, i.a.), and also on the *concept of medicine* and its subordinate concepts (those of diagnosis and prognosis, i.a.). And there would be such a body of writings on general *principles* of medicine, logical and ethical. These core writings on medicine, like the Babylonian Talmud, would incorporate unresolved disputes before the ideas' convergence to singularity, like that represented by the *Misneh Torah* of Malmonides (Section 2.1 above).

As an important supplement to all of this general theory of medicine, there would also be an understanding of the nature of a doctor's counterpart of a rabbi's personal, lifelong endeavor to fulfill his purpose as a knowing creature, of his quest for self-improvement as a spiritual teacher-by-example. That is, there would be a body of principles of *how students of medicine and medical doctors can best pursue the highest possible level of professionalism* in their particular disciplines of medicine. (Cf. Section 1.5.) If all of this were to exist, future doctors' introduction to medicine could encompass, if not largely focus on, an exposition of those concepts and principles (and their associated terminology) – that *general theory* – of medicine.

Even though none of this is the case at present, nevertheless *needed now is an attempt at introduction to medicine in the context of commitment to the ideal implied by that paradigm.* This means, first off, an attempt at careful development of *propositions* as to what this theoretical content might be – for beginning students of medicine "not to contradict, nor to believe, but to weigh and consider," as counseled by (the philosopher) Francis Bacon about the purpose in any 'serious' reading (ref. below).

Reference: Bacon F. *The Essays or Counsels Civil and Moral* (Edited with an Introduction and Notes by Brian Vickers). New York: Oxford University Press, 1999. P. 134.

Advancement of propositions for such an introduction to medicine is the mission in this book (cf. Preface), accompanied by the dream that suitably Talmudic-type discourse among scholars of medicine will result in an ever-more-well-established corpus of the fundamental tenets of the dauntingly complex realm of modern medicine, tenets that would be shared across the various disciplines of modern medicine, present and future.

2.3   Insufficiency of the Paradigm

Maimonides' approach to the scholarly content of concern to the rabbinate I regard as paradigmatic about the *requisite level of erudition* in the introduction to any learned profession (Sections 2.1 and 2.2 above). But it is necessary to appreciate, also, that, different from introduction to law, whether religious or secular, in a suitably scholarly introduction to the aggregate of medical professions the need is to focus on the *learned way of thinking, specifically, about matters medical,* and that the generic topic of learned – correct – thinking is the subject of the scholarly field of *logic* within philosophy.

Relevant to an appropriate introduction to medicine are, for a start, logicians' teachings concerning *concepts* – in respect to their generic nature, their definition, their hierarchy, and their role in thinking. And closely related to these are their teachings about the importance of apposite *terminology* for the representation of concepts, so as to avoid inept terms impairing apprehension of the concepts they denote.

Medicine is, most broadly, directed to human *health* in the inclusive meaning of this term, which encompasses absence of health in the narrow meaning of this term; and in point of fact, the principal concern in medicine actually is *ill-health* or *illness*, meaning absence of health. For this reason, the most central concepts to be addressed in an introduction to medicine are the tightly interrelated ones of health and ill-health/illness. And in the conception of these entities in particular – with focus on illness – there should be no tolerance for loose thinking about the essence of these entities nor for untidy terms taken to be antonyms of 'health' in the narrow meaning of this term and, thus, synonymous with 'illness.'

Identification of illness as the broadest concern in medicine is in keeping with an important teaching of logicians: *Concepts need to be introduced in a logical order, starting with the broadest* – most general, most inclusive – one, the one with the greatest 'extension.' With illness, suitably defined, identified as the core concept concerning the objects of medicine, logic calls, first, for identification of the members – 'species' – of this 'genus' of concepts, that is, the concepts for which ill-health is the 'proximate genus,' species of illness such as disease, suitably defined.

These entities have '*properties*' that flow, logically, from their respective essences; and the concepts of these too need to be introduced, again in a logical sequence.

Having introduced the general concepts that are of concern from the vantage of each of the disciplines of medicine while in themselves extrinsic to medicine, there naturally is a need to supplement these concepts with ones concerning medicine itself, again in a logical order. While the implicit question orientationally addressed above was, *What is medicine about*? the corresponding core question now is: *What is medicine itself* (in the professional meaning of the term)? It is quite obvious that the proximate genus of medicine (of doctors' services, i.e.) – in the 'family' of genera with members such as rabbinical services – can be taken to be professional healthcare. But it is quite a challenge to identify and properly conceptualize the 'specific difference' that logically distinguishes healthcare in the domain of medicine from its alternatives in the healthcare genus of professional services.

That which distinguishes medicine from 'paramedical' healthcare (nursing, i.a.) – it has to do with what I call *gnosis* – is the proximate genus for the ones on the next lower level of concepts of medicine, with diagnosis naturally one of these. Etc.

The learned way of thinking in medicine presupposes not only tenable concepts of the objects (extrinsic) of medical thinking and of medicine itself; it also presupposes *principles* that specify correct deployment of these concepts in medical thinking. Some of these naturally are imperatives of logic – material/major rather than formal/minor logic – others ones of *ethics*. Orientation to both of these two species of principles of medicine needs to be cultivated in an appropriate introduction to medicine.

And ultimately this: As philosophers – in the parent discipline of logic and ethics – have quite unanimously taught, all decisions about action/inaction are motivated by the decision-maker's pursuit of his/her own, personal happiness, those of Albert Schweitzer and Mother Teresa, for example. (The distinction between egoists and altruists lies elsewhere: in what it is about peoples' actions that makes them happier, whether serving themselves alone or acting out of concern for all who are affected by their chosen actions.) Hence, an introduction of students to medicine should also guide them in their pursuit of *professional happiness in medicine*. These precepts, I suggest, can be deduced from what I take to be the essence of medicine (Section 4.1.1) jointly with what I regard as the core principle of medical ethics (Section 6.3).

2.4 Significance of the Paradigm

2.4.1 Significance re Content

The paradigm in Section 2.1 supports the philosophy in the Preface about the way students preparing for careers in the practice of medicine need to be introduced to the field (Section 2.2). But: Is that orientational philosophy in need of support from some lateral, non-medical field of knowledge-based professional practice? Otherwise put: Isn't philosophically well-founded introduction to medicine obviously the one to embrace and, therefore, the basis of its contents already? That is, doesn't any serious teaching of medicine already begin with critical elaboration of fundamental concepts such as were touched upon in Section 2.3 above, etc.?

Instructive about the significance of that paradigm, insofar as it indeed supports the orientational philosophy in the Preface, is examination of the way students at Harvard Medical School have recently been introduced to medicine. The introduction, and the rest of the medical education too, at Harvard just as elsewhere, has been *devoid of foundation on any critical philosophy of medicine*, and both the content and the process of the education therefore have continued to be in considerable flux. A rather recent change in the education at Harvard was so dramatic that the university's president described it as "'Harvard's most important innovation of the 1980's'" (ref. below, p. 85):

Reference: Good BJ, DelVecchio Good M-J. "Learning medicine." The constructing of medical knowledge at Harvard Medical School. *In*: Lindenbaum S, Lock M (Editors). *Knowledge, Power, and Practice. The Anthropology of Medicine and Everyday Life*. Berkeley: University of California Press, 1993. (pp. 85–105.)

In this innovative teaching, "medicine is introduced as a science" (p. 89); and "it is made profoundly clear that learning medicine in the first 2 years is above all learning the biomedical sciences" (pp. 89–90).

In these studies, "a huge vocabulary is to be learned, a working vocabulary as large as most foreign languages, and competence in medicine depends on learning to speak and read this language. ... The language learned [is that of] a biochemical world, a world of cell biology and physiological systems as well as of discrete diseases" (pp. 97–98).

This innovation was dubbed "The New Pathway in General Medical Education" (p. 85). It was "designed to bring two new perspectives to bear on medical education: a new educational philosophy, focusing on tutorial-based active learning, a case method akin to that of the Harvard Business School, and new integrated curricular blocks; and a new vision of the physician for the twenty-first century – one who ... " (p. 85). It thus was, principally at least, an innovation of the process, rather than of the content, of the education.

Regarding, specifically, *introduction* to medicine at large in the framework of this "case method" of "active learning," the authors describe the first session on the first day of Week 1 for one of the groups of beginning students of medicine: After some organizational remarks, "the head of the course ... commenced with 'Okay, let's talk some science!'" (p. 89). The talking that ensued was about the death of a Bulgarian intellectual who had defected to Britain. "The discussions, directed by the students with selective guidance from the tutor, began with three basic categories written on the blackboard: Data, Hypothesis, Agenda" (p. 87). The authors quote one of the students subsequently remarking how the work on this first case, already, made them think of themselves as scientists (p. 89). (Cf. Section 1.5.)

That Bulgarian intellectual had been assassinated. This presumably was the result of his having done what intellectuals so commonly do: speaking truth to power. When still alive, he presumably would have been able, and willing too, to *speak truthfully* to the powers at Harvard about his own and other intellectuals' thinking about and experiences with doctors' services. He could have described how the intellectuals he has personally known, just like he himself, have not, as a matter of an a-priori truism, taken doctors' services to be science, nor have they found the theoretical framework and knowledge-base of these services to evidently be scientific in character; and in particular, never have they been advised by their doctors about their health problems to have been ones of cell biology and the like, indentified and also otherwise successfully dealt with on the basis of the doctors' mastery of such sciences (so-called 'basic' medical sciences; Section 1.4.1).

. . .

This glimpse into what passes as a modern academic conception of medicine – and studies of it sans introduction to the field – illustrates the significance of the paradigm adopted here (Section 2.1) as to the outlook in medical education it calls for (Section 2.2 and Preface).

Emblematic about the paradigm's significance on the *content* of the education is the change it calls for in the opening exclamation quoted above from an avant-garde program of medical education. For, the minimally yet dramatically edited version of that exclamation, called for by the paradigm, would be this: "Okay, let's talk some sense (*sic*)!"

The ensuing sensible talk might best focus on *the unintended consequences of the great accomplishments of laboratory-based medical science* (Section 1.4), most notably the common yet profoundly false and seriously education-distorting notion (in evidence above) that modern medicine is science and modern doctors are scientists (which they affect by donning scientists' laboratory coats). For, the paradigm calls for the teachers' – and secondarily the students' – commitment to certain *unassailable fundamental tenets* and all that rationally flows from these. And to be emphasized here is the overarching one of these tenets: *Even scientific medicine is not science; it is the aggregate of its constituent arts* (*sic*; Sections 4.1.1 and 4.1.5) of professional healthcare to people. (The logo of Harvard spells *Veritas*.)

2.4.2 Significance re Process

That 'case method' of learning reflects, directly, a report published in 1984 by the Association of American Medical Colleges; but indirectly it reflects and closely resembles its precursor – the *'problem-based learning'* that was initiated by Canada's McMaster University School of Medicine two decades earlier and is now in application in many medical schools around the world.

Wikipedia says that "Prominent educational researchers have reviewed what they describe as research studies using quality methodologies and concluded that in general project based approaches are less effective than direct instruction [refs.]," and that "studies supporting problem-based approaches are often methodologically biased."

I say that efforts at assuring effective teaching of medical students are not really timely to engage in before it is clear what modules actually belong in the medical common (Preface) and what the canonical contents of them should be taken to be. And as for *false tenets* such as the overarching element in that Harvard introduction to medicine – the notion that medicine is a science – *their ineffective teaching is preferable* to their effective, misleading teaching.

Harking back to the *paradigm* (Section 2.1) in this *process* regard as well, it can be said that in the education of rabbis, cultivation of effectiveness in teaching the Talmud beyond the canonical contents of the Torah became a well-justified concern only once the inconsistencies and disputes in the Babylonian Talmud had been settled, seemingly at least, in the *Misneh Torah*.

Part II
General Concepts of Medicine

Chapter 3
General Concepts of the Objects of Medicine

Contents

© Springer International Publishing Switzerland 2015
O.S. Miettinen, *Medicine as a Scholarly Field: An Introduction*,
DOI 10.1007/978-3-319-19012-9_3

3.0   Abstract

The *concept* of a thing – entity, quality/quantity, or relation – is the essence of the thing: that which is true of each instance of the thing and unique to it. Possession of concepts is a prerequisite for any thinking, and shared concepts together with shared terminology are prerequisites for successful communication. *General* concepts of medicine are ones that are generic and thus not discipline-specific.

The most elementary general concepts of medicine have to do with things that are extrinsic to medicine itself; they have to do with the *objects* of medicine, the things to which medical thoughts and actions are directed, human health in the sense of lack thereof – ill-health/illness – being the most central one among these.

In medical academia, even, great *confusion* prevails about the general concepts of the objects of medicine, starting from those of health and illness already; and there consequently is great confusion also about the terminology pertaining to the concepts.

This confusion is subject to resolution, however.

3.1   Health

In this introduction to the general concepts of the objects of medicine I use as points of reference my two dictionaries of medicine (refs. 1 & 2 below):

References:

1. *Dorland's Illustrated Medical Dictionary*. 28th edition. Philadelphia: W.B. Saunders Company, 1994.
2. *Stedman's Medical Dictionary*. 26th edition. Illustrated in Color. Baltimore: Williams and Wilkins, 1995.

The former is described (in its Preface) as "the outstanding authoritative guide to the language and usage of medicine and related health care fields." The latter is described, correspondingly, as "the most up-to-date authority on medical language" (Preface).

Notable about these dictionaries' self-descriptions is not only that they both claim authority but, also, that they take the purpose of a medical dictionary to be exposition of medical "language" – in English, specifically – rather than specification of the concepts to which the various medical terms refer and/or should refer. And as for the terms they address, neither one of these dictionaries specifies whether the intent is to be merely descriptive of the prevailing usage ('structuralist,' i.e.) or prescriptive/normative about it (specifying correct usage, i.e.), or perhaps both. As will become evident, this centrally relevant lexicographic duality is nowhere manifest in those two dictionaries; and very notably, the two generally are quite discordant mutually – nowhere acknowledging the existence of the other "authority" – and even internally inconsistent in what they say – commonly in an undisciplined, pleonastic way.

In an introduction to medicine already, the student should use whatever relevant readings the way all of us were taught by Francis Bacon to use them (ref. in Section 2.2): "Read not to contradict, nor to believe, but to weigh and consider." I read this counsel long ago, weighed and considered it, and came to the view that, as for scholarly writings, one indeed should not reflexively contradict, nor meekly believe, but should critically weigh and consider the content – thereby arriving at one's own, considered and defensible view about the matter at issue. And I suggest that this is the way the present introduction to medicine is most appropriately read (cf. Preface and Section 2.2). Authoritarianism I do not see to have any justifiable place in scholarly contexts (cf. above).

· · ·

From one of those two dictionaries of medicine one can read that *health* is:

a state of optimal physical, mental, and social well-being, and not merely the absence of disease or infirmity [ref. 1];

but according to the other one the term has three denotations:

1. The state of the organism when it functions optimally without evidence of disease or abnormality. 2. A state of dynamic balance in which an individual's or group's capacity to cope with all of the circumstances of living is at an optimal level. 3. A state characterized by anatomical, physiological, and psychological integrity, ability to perform personally valued family, work, and community roles; ability to deal with physical, biological, psychological and social stress; and freedom from the risk of disease and untimely death [ref. 2].

Before any critical weighing and considering of these definitions – or any of the others that will follow – the student needs to be clear on the *form of the logical definition* of the concept of a thing: it specifies the thing's 'proximate genus' and its 'specific difference' from all other species of things in that genus (ref. 3 below). For example, a triangle is a polygon (proximate genus) with three sides (difference distinguishing triangle from all other polygons).
     Reference 3: McCall RJ. *Basic Logic. The Fundamental Principles of Formal Deductive Reasoning.* 2nd edition. New York: Barnes & Noble, Inc., 1952.
     An added orientational principle naturally has to do with the *content* of the (formally correct) definition of the concept of the thing that a given term – word or group of words (ref. 3 above) – denotes. It seems to me that a *tenable definition* of a concept of medicine is one that can be agreed upon by practically all of those who are entitled to offering an opinion on the matter – upon their having duly weighed and considered the definition in question and the rationale behind it.
     In these terms, then, the first question is about the proximate genus of (human) health, of an individual's (rather than a population's) health. The two definitions above are in agreement that, as its proximate genus, health is a state of being, and this I believe to be agreeable to qualified opiners on the matter in general. Those two definitions are solely about the state of a person as a whole, but the medical concept of health unquestionably can be specific to a particular organ or system –

the cardiopulmonary system, for example. And indeed, the medical concept of a person's overall health is that of the aggregate of such site-specific states of health – in 'periodic health examinations,' for example.

What, then, is the specific difference of the healthy state from any unhealthy state of the cardiopulmonary system, to focus on this example? It definitely is not this system's "optimal physical, mental, and social well-being," nor is it its functioning "optimally without evidence of disease or abnormality." For, this organ system, like whatever other segment of the body, never has any level of mental or social well-being, and its functioning is optimal only in the competition performances of some top athletes (although such athletes, even, commonly find their cardiopulmonary functioning to have been suboptimal in particular competitions).

In truth, I dare say in contradiction to what those "authoritative" dictionaries of medicine say, the medical concept of health is, simply, *the state of a person's soma characterized by the absence of any ill-health/illness*, whether in reference to a particular organ or system or the soma at large – the concept of human soma being that of the body as distinct from the mind or the 'soul' and, in some conceptions, according to my dictionaries of medicine (refs. 1 & 2 above), exclusive of the germ cells.

As this is our first deliberately-developed definition of a medical concept, satisfying the principles governing the form and content of scholarly definitions (above), this is the appropriate context to introduce, also, that which a concept, according to logicians, in principle is and what its definition therefore is to express: the *concept* of a thing – health from the medical vantage in this instance – is the *essence* of that thing, meaning that which is true of every instance of that thing and unique to it (ref. 3 above). Thus, by the definition above, every instance of cardiopulmonary health is one of the absence of cardiopulmonary illness, and invariable – universal – absence of cardiopulmonary illness is unique to instances of health of the cardiopulmonary system.

A definition is meaningful only insofar as all of the '*notes*' (component concepts, explicit and also implied) in it are understood, as intended. In the present context this principally means the need to find, or else to develop, a tenable definition of the concept that the antonym of 'health' – 'ill-health' or '*illness*' – denotes.

. . .

The Larger Concept Before exploring the concept of illness as the somatic state representing the absence of health, it needs to be noted and underscored that the term 'health' in its common usage in medicine has a more inclusive denotation as well, not addressed by any of the three definitions above. *The inclusive concept of health subsumes illness*, along with health in the sense (narrow) in which health was addressed above. This awkward usage of the 'health' term persists because medicine lacks any more apposite term for the concept/genus whose constituent categories/species are health and illness (akin to 'gender' for the concept whose constituent categories are male and female).

As a prime example of this usage, *the common term 'health professions' denotes professions whose principal concern is ill-health, illness.*

Arguably at least, a better term for these professions would be *'illness professions'* – with the understanding that one possible state in respect to any given illness is its absence, that is, health in this respect. For, the concepts of the objects of medicine are not ones of health in the meaning of absence of illness; they are concepts of, and related to, illness. In line with this, fundamental to medicine are not normal anatomy and normal physiology but *patho*anatomy and *patho*physiology.

3.2 Illness

3.2.1 Illness as Somatic Anomaly

According to my Concise Oxford Dictionary (of Current English), the adjective 'ill' denotes the state of being "out of health," as in "taken ill with pneumonia." And the noun 'illness' it presents as denoting "a disease, ailment, or malady." My Oxford American Writer's Thesaurus specifies 'illness' as the antonym of 'health' (and, consistent with this, it gives 'health' as the antonym of 'illness'). Thus, in common conception, *illness is the absence of health*, consistent with the definition of health adopted in Section 3.1 above (different from what "authoritative" dictionaries of medicine say).

Here the need is to define illness directly, without reference to health. That thesaurus makes an orientationally relevant point by saying that "bad health" – illness, that is – is a "physical state," which means that illness is a *somatic anomaly*. With this the proximate genus of illness, it remains to understand the nature of the specific difference that in the somatic-anomaly genus is unique to illness (cf. Section 3.1 above).

Lung cancer is an example of somatic anomalies. This anomaly obviously constitutes an illness. And to be totally clear, it is the cancer per se that is the illness-definitional somatic anomaly in this example, not the symptoms and/or signs of it that the affected person may have and directly, immediately suffer from. This is to say that lung cancer is an illness even when it still is in the *latent*, 'pre-clinical' stage of its development, when the affected person still is "without evidence of disease or abnormality" (cf. Section 3.1 above). And an already-*overt*, 'clinical' case of an illness (manifest in symptoms and/or overt/clinical signs) is an illness even when it (temporarily) is in complete *remission*. These I believe to be the consensus views in medicine (again at variance with what is specified in an "authoritative" medical dictionary).

Two more examples of somatic anomalies as illnesses: deficient production of insulin (by the pancreas) and 'insulin resistance' (of tissues) are illnesses because of their (common) manifestations in symptoms and overt signs of 'diabetes.' (This term in its common usage is an abbreviation of 'diabetes mellitus,' where 'diabetes' denotes high throughput – here of water by the kidneys – and 'mellitus' refers to sweet taste – here to that of the urine. 'Diabetes' thus is not an apposite term for either one of those two illness-definitional anomalies.)

3.2.2   Illness as Hidden or Overt

The broader consensus-view illustrated by those examples I take to be this: A *hidden somatic anomaly* represents – is definitional to – an illness if it is, or potentially will be, manifest in symptoms and/or overt signs. But some clarification of this may be called for: A somatic anomaly such as lung cancer (in an intact person) remains hidden even if it has overt manifestations (which make for an 'overt case' of the hidden anomaly). Symptoms are inherently overt (to the patient), while a sign of a hidden anomaly derived from a diagnostic test is not an overt one.

What, then, about an *overt somatic anomaly*, specifically one not manifesting a deeper, hidden anomaly? Some types of overt somatic anomaly obviously are, in themselves, entities of ill-health/illness – various injuries and Down's syndrome, for example – while some other types of overt somatic anomaly are not viewed as representing illness – left-handedness and anomalously great body height, for example. An overt somatic anomaly, I suggest, represents an illness if, and only if, it shares with the overt manifestations of hidden somatic anomalies the quality of universal undesirability.

A distinction is to be made between anomalous somatic *states that have to do only with risk* for illness – osteoporosis, 'hypertension' (misnomer, as the concept is one of pressure, not tension), and immunodeficiency, for example – and those representing illness proper. Similarly, anomalous somatic states that are merely *intermediaries* in the connection between a hidden somatic anomaly and its manifestation in symptoms and/or overt signs *are not illnesses* – so that, for example, hyperglycemia is not the somatic anomaly definitional to either one of the two types of 'diabetes' (above).

In these terms, 'mental illness' is a misnomer, as the concept is not one of a (mentally manifest) somatic anomaly but, instead, that of the mental manifestations of a somatic anomaly (which remains unknown).

3.2.3   The Prevailing Confusion

Different from what I believe could be the medical consensus (above), according to my medical dictionaries (refs. 1 & 2 in Section 3.1 above) *illness* is:

> A condition marked by pronounced deviation from the normal healthy state; *sickness* [ref. 1; italics added];

or it is:

> An interruption, cessation, or disorder of body functions, systems, or organs [ref. 2],

with the latter source presenting not only 'sickness' but also '*disease*' and '*morbus*' as synonyms of 'illness.'

In these definitions I take note of, and underscore, the lack of distinction-making among the concepts denoted by those various terms. And rather than parsing these manifestations of confusion about the most elementary one of all of the general concepts of medicine, I proceed to critical examination of the concepts of sickness and disease, distinguishing these from the concept of illness.

3.3 Sickness

3.3.1 Sickness in Health

My medical dictionaries (refs. 1 & 2 in Section 3.1) define, under 'sickness,' dozens of entities with 'sickness' in the very terms denoting them. Examples of particular sicknesses from one of those sources (ref. 2) are these four:

> motion s., the *syndrome* of ... caused by ...;
> milk s., a *disease* caused by ... [whose] clinical manifestations include ...;
> green tobacco s., an *illness* of tobacco harvest workers ... characterized by ...; [and]
> decompression s., a *symptom complex* caused by ... [and] characterized by ....

It may be noted from these definitions of particular sicknesses that sickness is not given as the proximate genus of any of these – as it should (Section 3.1), in lieu of disease and illness, for example. The same is true of all of the other definitions of particular sicknesses in this dictionary.

Let us examine *motion sickness* more closely, as an instructive example. In its detailed definition that same dictionary specifies it as:

> the syndrome of pallor, nausea, weakness, and malaise, which may progress to vomiting and incapacitation, caused by stimulation of the semicircular canals during travel or motion as on a boat, plane, train, car, swing, or rotating amusement ride.

The other dictionary (ref. 1), in turn, defines this entity as:

> Sickness [cf. remark above] caused by motion experienced in any kind of travel, such as sea sickness, train sickness, car sickness, and air sickness.

Evidently, thus, motion sickness is not thought of as a somatic anomaly (of a specified type). It thus is not thought of as an illness (cf. Section 3.2.1). Nor is it thought of as a manifestation – overt – of an underlying somatic anomaly. Motion sickness thus is an example *sickness in health*, in the absence of illness. Instead of a somatic anomaly, the proximal cause of motion sickness is a circumstantial one and, thus, extrinsic to the person's soma per se.

The other three sicknesses above are, in their respective conceptions, analogous to motion sickness. In decompression sickness the extrinsic proximal cause (impinging on a healthy soma) is exposure to low or rapidly lowering ambient pressure (among divers, mainly), while in the other two it is exposure to a poison. In green tobacco sickness the poison is nicotine from exposure to wet, green tobacco leaves (among harvesters of tobacco, mainly). In 'milk sickness' (misnomer) the poison is another type of plants-produced toxin (to which exposure is conveyed by ingested milk, milk products, or meat from cattle or sheep with a condition called trembles from this type of toxin). Sickness from imbibing an alcoholic beverage in excess is akin to these last two examples; it ('hangover') could be termed libation sickness. From the essence of sickness in health flows the property of that it is not only initiated by also sustained by its extrinsic proximal cause – the removal of which provides for recovery.

For any particular sickness defined by its extrinsic direct cause, the definition must not include any attempted specification of what that sickness actually is; for, the sickness from a given cause generally is quite variable. Thus, motion sickness is to be defined – different from the first definition above – as: sickness with motion as its proximal cause. This sickness per se is some combination of symptoms and signs such as those given in this condition's first definition above.

A sickness in health can also have an intrinsic direct cause, a normal, transient somatic state (hormonal, say).

3.3.2 Sickness from Illness

While a person can be sick from a direct cause of it that is extrasomatic (environmental or behavioral; Section 3.3.1 above), or from an exceptional but normal somatic state (hormonal, say), the direct cause of sickness usually is a somatic anomaly.

Sickness is the essence of the suffering from a *hidden* somatic anomaly; for it – the aggregate of symptoms and/or overt signs – is the terms in which the affected person experiences the presence of such an anomaly. But the direct cause of sickness from a hidden illness-definitional somatic anomaly may not be that anomaly itself; it may be a hidden somatic consequence of that basic anomaly. For example, sickness from 'diabetes' (Section 3.2.2) commonly results from an abnormal level of blood sugar (hyperglycemia) as its direct cause, rather than from the anomaly in insulin production or in tissue response to it that is the essence of this illness and the cause underlying the direct cause of the sickness.

What, then, about illnesses whose essence is an *overt* somatic anomaly? Some overt anomalies of the soma do cause sickness, directly. For example, some lesions of the skin cause itching, and pain is quite ubiquitous a consequence of overt injury (the active process of this, distinct from its outcome state).

But there are overt anomalies of the soma from which the affected person suffers directly, rather than through the mediary of sickness from the illness. Thus, one does not think of sickness from – of symptoms and/or overt signs of – vitiligo of the skin or overt malformations, for example.

Overall, thus, when a person is suffering from a sickness, this has a direct cause which, when not extrasomatic, is either an illness per se or a hidden anomaly caused by the illness.

3.3.3 Symptoms and Signs

While the concept of illness (Section 3.2) involves the subconcept of hidden somatic anomaly and its potential manifestation in symptoms and/or overt signs, the concepts of sickness in health (Section 3.3.1) and sickness from illness (Section 3.3.2 above) involve symptoms and/or overt signs in a more central role: they are the *essence of sickness* – present in every instance of sickness, whether from an intrinsic or extrinsic direct cause. The concepts of symptom and sign thus are of central concern in medicine, as central as the concept of illness.

My dictionaries of medicine (refs. 1 & 2 in Section 3.1) define *symptom* as:

any subjective evidence of disease or of a patient's condition, i.e., such evidence as perceived by the patient; a noticeable change in the patient's condition indicative of some bodily or mental state [ref. 1];

and alternatively as:

Any morbid phenomenon or departure from the normal in structure, function, or sensation, experienced by the patient and indicative of disease [ref. 2].

And as may have become almost expected, I have a corresponding third opinion: As I see it, the concept of symptom in medicine is that of a *sensation* (e.g., a pain or nausea) indicative of the presence of some underlying problem – either somatic anomaly (illness-definitional) or an extrasomatic stressor (causal to sickness in health). It is subjective in the meaning of being perceptible to the affected person alone.

Closely related to the concept of symptom is that of *sign*. Those dictionaries of medicine define sign as:

an indication of the existence of something; any objective evidence of a disease, i.e., such evidence as is perceptible to the examining physician, as opposed to the subjective sensations (symptoms) of the patient [ref. 1];

and alternatively as:

Any abnormality indicative of disease, discoverable on examination of the patient; an objective symptom of disease, in contrast to a symptom that is a subjective s. [symptom] of disease [ref. 2].

Again, I think that the appropriate definition is different from both of those. The concept of sign in medicine is, I think, simply the *objective* counterpart of symptom. It is commonly thought of as inherently being a finding (abnormal) from the doctor's direct examination (physical or other, e.g., electrocardiographic) of the person; but laboratory findings (abnormal) should be seen to be signs just the same. A sign is objective in the sense that it would be agreed upon by (practically) all qualified observers.

3.3.4 Syndrome of Sickness

Closely related to the concepts of symptom and sign is that of *syndrome*. My medical dictionaries define it as:

> a set of symptoms which occur together; the sum of signs of any morbid state; a symptom complex [ref. 1];

and alternatively as:

> The aggregate of signs and symptoms constituting together the picture of the disease [ref. 2].

These definitions miss the essence of the aggregates of overt problems that in medicine are properly – with specific meaning – termed syndromes. The essence of syndrome is illustrated by and learnable from those patterns that have classically – and of necessity – been called syndromes, just as the essence of sickness is illustrated by and learnable from entities that have 'sickness' in their names (cf. Section 3.3.1).

A widely familiar and instructive example of syndromes is the major congenital anomaly commonly termed Down's syndrome, which is a particular aggregate of findings from physical examination of the patient. This aggregate of findings was taken to be due to some particular type of deeper congenital anomaly well before the actual nature of this hidden anomaly (trisomy, of chromosome 21 in most cases) was discovered. Before this discovery, the illness had to be labelled on the basis of this syndrome of clinical findings, while now its appellation could be based on its congenital cause – as the *trisomy syndrome*, say.

Another instructive example, quite recent in its occurrence (transient, epidemic), is phocomelia, a congenital condition defined by an equally distinct set of findings from physical examination. Its cause (congenital) – also unique a priori – turned out to be mother's use of a novel medication, thalidomide, as a remedy for 'morning sickness' from early pregnancy. The aggregate of the features of phocomelia could be called the *thalidomide syndrome*.

A well-familiar acquired counterpart of these anomalies is Cushing's syndrome, of 'moon face' etc., caused by excess production of cortisol (due to neoplasm of the adrenal cortex or of the anterior lobe of the pituitary gland) or by the use of a cortisol medication. This constellation of (overt) findings could be called the *cortisol syndrome*.

As these three examples suggest and illustrate, the classical concept of syndrome in medicine is an aggregate of overt features such that *it, by its recurrent singular nature, is indicative of a singular type of proximal cause*. In the jargon of medicine, a syndrome in this meaning of the term is 'pathognomonic' of its direct cause, possibly congenital or else the presence of a specific underlying somatic anomaly or a specific extrasomatic cause, with the concept useful on this account. In those first two examples the syndrome is not an aggregate of symptoms and/or signs, of sickness manifesting the presence of its proximal cause (which was prenatal); but in the third example the syndrome is one of sickness, representing overt manifestations of its direct cause (somatic or extrinsic).

Genuine syndromes are quite uncommon, ones of sickness in particular. But there has recently been an inflation in the use of the 'syndrome' term – and an obfuscation of the concept in consequence of this all the way to essential meaninglessness. One eminent example of these neologisms is 'acute coronary syndrome,' which is not a singular set of symptoms and/or overt signs diagnostic of acute coronary heart disease. The same is true of the 'acquired immunodeficiency syndrome,' AIDS, as a manifestation of its underlying virus (HIV) infection. Particularly muddled is the now-eminent concept of 'metabolic syndrome,' as it is variously defined. According to the NIH, it is constituted by having "at least three" (*sic*) of these five conditions: large waistline, high triglyceride level, low HDL cholesterol level, high blood pressure, and high fasting blood sugar. (Only one of these five is a symptom or an overt sign, and there is no conception of a singular underlying anomaly at the root of this 'syndrome.')

3.3.5   Overview

All in all, thus, sickness can be (and most commonly is) a *complex of symptoms and/or overt signs of the presence of a hidden somatic anomaly* (illness); but it can also be *such a complex without a somatic anomaly* as the proximal cause of it. In the latter case the proximal cause of the sickness is a circumstantial one, generally an environmental or behavioral stressor but sometimes a normal somatic situation (such as pregnancy as the proximal cause of 'morning sickness'). Such a complex *need not be a syndrome.* Nor is sickness inherently a complex of elements (from which the affected person directly suffers); *it can be an isolated problem*: a pain, a fever, a skin rash, a 'hot flash' (from menopause), etc., occurring alone.

Sickness is a concern in medicine as central as illness.

As a closing note on this important and rather challenging topic, I hark back to the concept of 'health professions' at the end of Section 3.1: *When not concerned with illness per se nor with sickness from illness, the principal concern of the health professions is sickness not due to illness* – sickness from the use of medications, most notably.

3.4   Disease

According to my dictionaries of medicine, an element in the concept of health is absence of "disease or infirmity," or of evidence of "disease or abnormality" (Section 3.1); and symptom they characterize as evidence of "disease or of a patient's condition," or as indicative of "disease" (Section 3.3.3). According to these dictionaries, thus, the closest candidate for being the antonym of 'health' appears to be 'disease,' while the critical weighing and considering (à la Francis Bacon, Section 2.2) in Sections 3.1 and 3.2.1 led to the view that absence of health is tantamount to presence of illness.

This impression that the absence of health is taken to be tantamount to the presence of disease prompts the question: Is 'disease' an apt synonym of 'illness'? In other words, the question here is this: Is disease, like illness, *any* somatic anomaly at least potentially causal to suffering? I hold, firmly, that this is not the case; that in its actual medical conception, *disease is a particular species of illness*. The Latin word for it is *morbus*, and the denotation of 'disease'/'morbus' in careful medical usage is not simply absence of health; it is more specific than this.

Somatic *defect* can be an illness, and so can *injury* to the soma, as both of these are somatic anomalies potentially (or actually) causal to suffering. And while they thus can be illnesses, they are not thought of as potentially being diseases. For these two species of illness the respective Latin words are *vitium* and *trauma* (not 'morbus').

A somatic defect is a more-or-less stable, state-type anomaly, whereas a disease is defined by a *process*-type somatic anomaly – 'disease process' – evolving over time. Examples of somatic defect constituting an illness (though not a disease) are the two types of anomaly underlying the sickness in 'diabetes' (Section 3.2.1), while lung cancer is an example of disease-definitional somatic process. Accordingly, in reference to somatic anomaly one does not speak of 'mental disease' but only, non-specifically, of 'mental illness,' as the somatic anomaly underlying the sickness is not known to be of the process type (while in reference to the sickness, 'mental illness' is a misnomer; Section 3.2.2).

An injury, like a disease, is of the process type; but it nevertheless is not a disease. Injuries differ from diseases in the nature of their genesis – pathogenesis and etiogenesis (Sections 3.5 and 3.6 below).

3.5 Pathogenesis

Pathogenesis of illness is an eminent topic in *pathology*. My dictionaries of medicine (refs. 1 & 2 in Section 3.1) define this field as:

> 1. That branch of medicine which treats of the essential nature of disease, especially of the structural and functional changes in tissues and organs of the body which cause or are caused by disease. 2. The structural and functional manifestations of disease [ref. 1];

and alternatively as:

> The medical science, and especially practice, concerned with all aspects of disease, but with special reference to the essential nature, causes and development of abnormal conditions, as well as the structural and functional changes that result from the disease process [ref. 2].

And the respective definitions of *genesis* are: "the coming into being of something; the process of originating" (ref. 1) and "An origin or beginning process" (ref. 2).

For *pathogenesis*, then, the definitions in those two sources are:

the development of morbid conditions or of disease; more specifically the cellular events and reactions and the other mechanisms occurring in the development of disease. drug p., the production of symptoms by the use of drugs [ref. 1],

and

The pathologic, physiologic, or biochemical mechanisms resulting in the development of a disease or morbid process [ref. 2],

respectively. An eminent species of pathogenesis is *oncogenesis*. For it the definitions are given as "the production or causation of tumors" (ref. 1) and as "Origin and growth of a neoplasm" (ref. 2). Closely related to oncogenesis is carcinogenesis, defined as "the production of carcinoma" (ref. 1) and as "the origin or production, or development of cancer, including carcinoma and other malignant neoplasms" (ref. 2).

I suggest, for a start, that *genesis* has to do with something that has come into being; that it is a *retrospective* concept from the vantage of an existing reality – the universe as it is, or a person's existing case of a somatic anomaly, for example. The genesis of an existing reality is the way this reality came, or generally comes, into being. The genesis of whatever exists has two very distinct aspects: the descriptive – *process* – aspect of how and the explanatory – *causal* – aspect of why.

As is apparent from those two definitions of *pathogenesis*, this is the *descriptive*, acausal or process, aspect of a somatic anomaly's having come, or generally coming, into being, distinct from the explanatory or causal aspect – the anomaly's etiogenesis (Section 3.6 below). The pathogenesis of a somatic anomaly, such as the process of changes leading from normal tissue to a malignancy in carcinogenesis, is commonly characterized as being a matter of "mechanisms" (as in both of the definitions above), but the concept of this in reference to successive stages of an anomaly's (incipient cancer's, say) having come or generally coming into being I find to be quite obscure.

The concept of *mechanism* according to those dictionaries of medicine is:

1. a machine or machine-like structure. 2. the manner of combination of parts, processes, etc., which subserve a common function. 3. the theory that the phenomena of life are based on the same physical and chemical laws as in the inorganic world; opposed to vitalism

according to one of them (ref. 1) and

1. An arrangement or grouping of parts of anything that has a definite action. 2. The means by which an effect is obtained

according to the other (ref. 2). I cannot see the involvement of anything like this to be inherent in the pathogenesis of a somatic anomaly (definitional to an illness).

Illustrative of the concept of pathogenesis is the way in which a *sequela* of a disease (or an injury) came (or comes) into being, myocardial scar as a sequela of the process of myocardial infarction or cirrhosis of the liver as a sequela of the process of hepatitis, for example. The disease (or injury) process is the pathogenesis of any sequela of it. In the inherently descriptive terms of pathogenesis, the somatic process ending in a sequela is thought of as having proceeded (or to generally proceed) from one stage to the next, ultimately coming to its completion and, at this point, leaving the sequela as a vestige of it. Explanation of neither the inception nor the progression of this process is involved in the concept of the pathogenesis of a sequela of an illness (or of the illness itself). And insofar as explanation is a concern, it is about causation of the changes involved (cf. above), not 'mechanisms' in these.

In Section 3.4 above I made the point that while an *injury* shares with disease the process-type essence of the somatic anomaly, it differs from disease in terms of pathogenesis. The somatic anomaly that is definitional to a disease has its genesis in terms that are intrinsic to biology, pathobiology to be specific. In contrast to this, the genesis of an injury involves a non-biological stress that is more than the soma can bear, resulting in somatic disintegration. The stress can result from movement of the body (as in tennis injuries, i.a.), or its source can be extrinsic to the body (as in burns, frost bites, bullet wounds, and bed sores, i.a.).

The pathogenesis of congenital malformations, taking place in embryonic organogenesis, bears no resemblance to that of diseases or injuries.

3.6   Etiogenesis

3.6.1   The Concept of Etiogenesis

The 'etiogenesis' term – a neologism of mine – cannot be found in my dictionaries of medicine. It denotes what 'etiology' also should be understood to denote.

My dictionaries of medicine (refs. 1 & 2 in Section 3.1) define *etiology* as:

> the study or theory of the factors that cause disease and the method of their introduction to the host; the causes or origin of a disease or disorder [ref. 1];

and as:

> 1. The science and study of the causes of disease and their mode of operation. Cf. pathogenesis. 2. The science of causes, causality; in common usage, cause [ref. 2].

Etiology of illness is the focus of modern epidemiological research on the population level; and the International Epidemiological Association defines this concept of its central concern in its dictionary (ref. 1 below) as:

> Literally, the science of causes, causality; in common usage, cause.

(Cf. ref. 2 above.)

Reference 1: Porta M (Editor), Greenland S, Last J M (Associate Editors). *A Dictionary of Epidemiology*. 5th edition. Oxford: Oxford University Press, 2008.

The topic of causality/causation has been a controversial one among philosophers, even as to its very admissibility in the framework of the principle of parsimony (which calls for elimination of unnecessary, superfluous concepts, especially ones that are not ontologically 'real'/admissible). Hume argued that, instead of causality, observable – and therefore learnable about – are only "constant conjunctions," and he thus had no use for the concept of causality. Kant agreed with Hume's premise – that causation is not a phenomenon, and that it therefore is not observable – but he did not agree that the concept of causality thereby is useless or even inadmissible. Kant held that while a concept generally derives from observations on the thing – entity, quality/quantity, or relation – at issue, some concepts are innate to the human mind, ones that in this sense are concepts a priori. To Kant, a prime example of this was causation.

In medicine the concept of causation is of absolutely central relevance. Without causal thinking medicine would be as passive about – and hence as inconsequential to – human health as cosmology is about the goings-on in the cosmos – 'life' cycles of stars, for example. And this would be profoundly antithetical to medicine as an aggregate of 'healing' professions, in the active meaning of the professionals causing (sic) illnesses to heal. (This is commonly said to be in the essence of medicine, but it rarely is true even of modern medicine.)

Causal concerns in medicine are of two very different generic types. One of these is about causality in the passive sense of causal understanding and explanation of phenomena of health that present themselves – most notably cases of illness or of sickness not due to illness, but of somatic/constitutional risk-factors as well. This is the etiogenesis/etiology aspect of the somatic anomaly definitional to an illness, or of sickness not manifesting a somatic anomaly (Section 3.5 above), or of somatic risk-factor status. This passive – retrospective – concept of causation is the concern here, while causation in the active sense of intended, prospective effects (and unintended side-effects) of medical interventions, and of health-related choices of lifestyle, are addressed in Sections 4.4 and 4.6.

So, harking back to the dictionary-definitions of etiology above, is the proximate genus (Sections 2.3 and 3.1) of etiology/etiogenesis "study or theory," or "science," or "cause"? No, it is none of these. For the concept of etiology/etiogenesis in medicine the proximate genus is genesis (Section 3.5 above) of an entity of health; and within this genus the specific difference definitional to etiology/etiogenesis is the causal/explanatory aspect of this (as distinct from the acausal/descriptive one; Section 3.5 above).

In principle, as a Kantian-type apriorism (above), the causal origin of a case of an entity of health always is its antecedent presence – and action – of a sufficient cause of the case coming into being. In medicine, a sufficient cause commonly is some aggregate of insufficient causes – constitutional (genetic, other congenital, and/or 'acquired'), environmental, and/or behavioral. For example, a case of myocardial infarction always is – as a matter of an aprioristic way of thinking – an obligate consequence of its antecedent sufficient cause, generally some combination of constitutional factors (coronary, hematologic) possibly together with some (stress-inducing) behavioral and, even, environmental ones.

3.6.2   Concepts of Causation

My dictionaries of medicine define *cause* as:

> that which brings about any condition or produces any effect [ref. 1 in Section 3.1];

and as:

> That which produces an effect or condition; that by which a morbid change or disease is brought about [ref. 2 in Section 3.1].

As I critically weigh and consider these definitions (à la Francis Bacon, Section 2.2), I perceive a lack of specificity to etiogenetic/etiologic causation, to the essence of this particular species of cause in medicine.

As I noted above, etiogenetic cause generally is thought of as *an element in some sufficient cause(s)* of the case of the entity of health in question, as an antecedent that in itself generally is an insufficient cause of such a case. And when at issue is a somatic anomaly (rather than sickness not due to illness), a cause of it is the antecedent category in one of the three principal dimensions of these (Section 3.6.1 above). Finally, any such category never is a cause in itself but only relative to a particular *alternative* to it, in the same subdimension. For example, for a case of myocardial infarction, its immediately antecedent physical exertion (in the behavioral dimension) of the patient could have been a cause of it only in terms of some particular lesser level of exertion as the alternative. Thus, an etiogenetic cause is a category of a *risk factor* – causal indicator of risk – relative to a particular other category of this, such as a given level of physical exertion in that example relative to a particular lower level of this risk factor.

In these terms, the medical concept of *etiogenetic causation* – akin to that of liability in tort law – is this: the event of health would not have occurred (at the time it did though potentially later) but for its antecedent presence of the cause at issue in lieu of its defined alternative, ceteris paribus (i.e., all else having been the same). For example, a history of a given level of 'hypertension' was a cause of a case of stroke (hemorrhagic) that has occurred, relative to a particular lower level of its antecedent blood pressure, if the stroke would not have occurred with that alternative antecedent (ceteris paribus).

. . .

Some *generic types of cause*, apparently in the meaning of etiogenetic cause, are defined in my dictionaries of medicine. Suffice it to consider only some of them in only one of these sources (ref. 1 in Section 3.1):

> primary c., the principal factor contributing to the production of a specific result;
> secondary c., one that is supplemental to the primary cause;
> predisposing c., anything that renders a person more liable to a specific condition without actually producing it;
> immediate c., a cause that is operative at the beginning of a specific effect; called also precipitating c.; [and]
> c. of death, the injury or disease responsible for death.

These specifications are quite similar to their counterparts in the other dictionary (ref. 2 in Section 3.1).

In critically weighing and considering these specifications, it is necessary to bear in mind the understanding (aprioristic) above, that *a generally insufficient cause* of a particular type of phenomenon can have been a sufficient cause of it in a particular case of this, and that it actually *was a sufficient cause insofar as it was a cause at all* (ceteris paribus) – by completing to sufficiency an otherwise insufficient aggregate of solely insufficient causes, while its alternative would not have had this role. It *either was or was not causal*, in this all-or-none sense of its role in the etiogenesis of a given case of the phenomenon.

Against this background of principles, let us first weigh and consider the duality purportedly constituted by the *"primary"* and a *"secondary"* cause – by the "principal" and a "supplemental" cause jointly involved in the causation of an event of health. Let us denote these two by A and B, respectively. Implied is that A without B would not have been enough to bring about the case at issue, and that A supplemented by B was a sufficient cause of the case. But similarly, B without A would have been insufficient, while B jointly with A was sufficient (ceteris paribus). There thus is no rational basis for ranking these two according to the extent to which they contributed to the etiogenesis of the case at issue. If B had a role at all, it was as critical as that of A (and conversely).

What about a *"predisposing"* cause in that meaning of increasing the susceptibility to the phenomenon in question without actually causing it? Osteoporosis is a predisposing cause of 'hip' fracture, meaning that its presence implies an increased probability that a fall, if it occurs, completes a sufficient cause of that injury. If the fracture results from a fall but only because of the person's degree of osteoporosis, then the predisposing condition, just as the fall, was a cause of it – alone an insufficient cause but in the context in question a sufficient one (cf. above). In general, any insufficient cause that was present before it became an element in a sufficient cause was a predisposing cause.

While said in that dictionary is that an *immediate* cause is one that is "operative at the beginning of a specific effect," it needs to be understood that a sufficient cause always is operative when a case of an event of health comes into being, and that all of the components jointly constituting the sufficient cause therefore are operative at that time. And also to be understood is that a synonym for 'immediate cause' is 'direct cause,' the respective antonyms of these being 'mediate cause' and 'indirect cause.' A mediate/indirect cause is a cause of an immediate/direct cause – of a constitutional, environmental, or behavioral element in the sufficient cause of a case of an illness that has come into being.

A point of note in this context is that *in the behavioral and environmental dimensions, only immediate/direct causes of phenomena of health are objects of etiogenetic thought in medicine*. Thus, when a particular use of a medication (as a behavior in lieu of its defined alternative behavior) is thought to have been causal to a case of stroke (hemorrhagic), causes of that behavior – those mediate/indirect causes of the case – are not thought of as having been etiogenetic to the case. Analogously, causes of physical exertion are not thought of as having been etiogenetic to myocardial infarction in the example above. And where exposure to pollution of ambient air (in the workplace, say) is thought of as an etiogenetic factor for a case of illness, causation of that pollution is not a matter of medicine. On the other hand, while atheromatosis is a cause (constitutional, acquired) of myocardial infarction and of stroke, causes of this cause – these mediate/indirect causes of these somatic anomalies – are of concern in medicine.

So, is a cause of a case of an illness, in the usual meaning of immediate/direct/proximal cause, a *precipitating* cause of the case, as per that dictionary? The sufficient cause of the case obviously is a cause that precipitated the occurrence of the case; but as for its solely insufficient components – all of them immediate/direct causes of the case (cf. above) – only the one that was the last entry into that illness-precipitating aggregate of causes can be said to be the component cause that precipitated the occurrence, while the others were predisposing causes. Physical exertion can have precipitated the occurrence of myocardial infarction in a person who constitutionally was susceptible to such precipitation.

Cause of death – of death that has occurred – is a common object of medical thought, but it is not a topic of ordinary etiogenetic thought in medicine, as death is not a somatic anomaly definitional to an illness nor an entity of sickness with an extrinsic direct cause. The concept of cause of death is not suitably defined by saying that it is "the injury or disease responsible for death." Causes of death are not all injuries or diseases; for they include suffocation, poisoning, and starvation, among others. And when the cause is said to be an injury or a disease, or a defect for that matter, the immediate/direct cause actually tends not to be that illness per se but a complication of it – ventricular fibrillation as a complication of myocardial infarction, or sepsis as a complication of pneumonia, for example.

3.6.3 Microbes as Pseudocauses

As a closing exercise on the important, challenging concepts of causation here, let us endeavor to understand the general nature of the (immediate/direct) *causation of tuberculosis*. This disease is now commonly seen to be universally caused by a microbial agent, the discovery of which was the result of work by Robert Koch (1843 – 1910) and the basis of his Nobel Prize, in 1905.

But, as described in a philosopher's treatise on the history of medicine (ref. 2 below), *the agent Koch discovered got to be seen as the universal cause of tuberculosis in consequence of Koch's definition (sic) of tuberculosis as infestation of the body with the bacillus he had discovered*:

> once Robert Koch had defined tuberculosis as infestation by the tubercle bacillus, he could deny that all other cases (which everyone *knew* to be tuberculosis) really were tuberculosis (namely, symptomatically and pathologically indistinguishable cases 'due to leprosy bacilli, brucella bacteria, histoplasma fungi, beryllium poisoning, or the unknown agent of sarcoidosis' [ref.] but lacking the tubercle bacillus). [pp. 52–53.]

Reference 2: Carter KC. *The Rise of Causal Concepts of Disease. Case Histories.* Burlington: Ashgate Publishing Company, 2003.

Now, any genuine cause of an illness always is an antecedent of it, and the (immediate/direct) cause of the body's infestation with the bacterium Koch discovered – Mycobacterium tuberculosis – actually and universally is its antecedent 'effective exposure' to the bacillus (environmental/behavioral component/insufficient cause) together with susceptibility to infection by this exposure (constitutional component/insufficient cause). *A lesion/anomaly defined as having a particular agent as an integral element of it is not caused by that agent*; its proximal cause in whichever given case is the aggregate of factors that brought about the agent's entry into the tissue. Thus the bacterium per se, which Koch took to be the universal cause of tuberculosis, is not the, nor even a, cause of this disease, as this disease was defined before Koch or even as Koch redefined it. (The bullet involved in the genesis of a bullet wound is not the, nor even a, cause of the injury. The immediate/direct/proximal cause of a bullet wound universally is exposure to the trajectory of the bullet's flight and susceptibility to the bullet's invasion into the body, due to lack of sufficient armor.)

My dictionaries of medicine, however, define *tuberculosis* in the spirit of Koch, as:

Any of the infectious diseases of man and animals caused [*sic*] by species of *Mycobacterium* and characterized by the formation of tubercles and caseous necrosis in the tissues [ref. 1 in Section 3.1];

and alternatively as:

A specific disease caused [*sic*] by the presence of *Mycobacterium tuberculosis*, ..., the anatomical lesion is tubercle, ... [ref. 2 in Section 3.1].

An apposite term for this disease, so long as that agent is to be in the essence of it, would be '*mycobacteriosis*,' in line with terms such as 'asbestosis' and 'trichinosis,' for example. Neither one of those dictionaries says that asbestosis is caused by asbestos fibers; they both say that it is caused by inhalation (*sic*) of asbestos fibers. And similarly, neither one of those dictionaries says that trichinosis is caused by the parasite; it is said to be "produced by eating undercooked meat containing ...," or "resulting from ingestion of raw or inadequately cooked pork ... that contains ..."

In line with that prevailing, Kochian conception of tuberculosis, measles is said to be "caused by a paramyxovirus"; poliomyelitis is said to be "caused by a virus, usually a poliovirus but occasionally by a coxachie virus or echovirus"; etc. (ref. 1 in Section 3.1); or: measles is said to be "caused by m. virus"; poliomyelitis (acute) is said to be "caused by the poliomyelitis virus"; etc. (ref. 2 in Section 3.1).

Orientational to all of this are the opening words in the philosopher's book alluded to above (ref. 2):

Of the numerous changes that have occurred in medical thinking over the last two centuries, none have been more consequential than the adoption of what Robert Koch called the etiological standpoint [ref.]. The etiological standpoint can be characterized as the belief that diseases are best controlled and understood by means of causes and, in particular, causes that are *natural* (that is, they depend on forces of nature as opposed to the wilful transgressions of moral or social norms), *universal* (that is, the same cause is common to every instance of a given disease), and *necessary* (that is, a disease does not occur in the absence of its cause) [ref.]. This way of conceiving disease ['universal' and 'necessary,' as defined here, actually are not distinct] has dominated medical thought for the last century. As new diseases like Legionnaires' disease or AIDS emerge, efforts to control and to understand them focus immediately on the quest for such causes. [P. 1.]

This statement has to do with *somatic anomalies as causes of sicknesses* and is true only of sicknesses of the syndrome type (à la certain type of neoplasm as the cause of Cushing's syndrome; Section 3.3.4). The mycobacteriosis anomaly, like communicable-disease anomalies quite generally, does not cause a syndrome-type sickness. Nevertheless, the idea of defining illnesses as particular types of somatic anomaly at the root of sicknesses (Section 3.2.1) has become well-established in medical thought and culture, even though its success still is incomplete, most notably in respect to illnesses manifest in mental sicknesses.

So well-established indeed is, already, the thinking about illnesses as commonly being hidden somatic anomalies causal to sicknesses (Section 3.2.2) that *the modern concept of etiology mainly is that of the causal origin – etiogenesis – of illness itself* – of the somatic anomaly definitional to it – and not the causal origin of sickness in its underlying illness. This has been the idea in this Section all along, though supplemented with etiology/etiogenesis of sickness due to non-somatic, circumstantial/extrinsic immediate/direct causes (and normal somatic causes such as pregnancy).

But those definitions of particular infectious diseases above illustrate how poorly understood the very concept of etiology/etiogenesis still is in the context of this very eminent species of disease, as the agent definitional to the somatic anomaly is not viewed as such – as a feature in the essence of the anomaly – but as the cause (*sic*) of the anomaly.

The conceptual underdevelopment of and confusion about the etiogenesis of infectious diseases merits some further illustration. One of my dictionaries of medicine (ref. 1 in Section 3.1) specifies the distinguishing quality of infectious diseases – their being infectious, that is – as their being "caused by or capable of being communicated by infection"; and infection it defines as: "1. invasion and multiplication of microorganisms in body tissues … 2. an infectious disease." In these terms, tuberculosis – this infectious disease – is infectious because it is caused by invasion and multiplication of M. tuberculosis in body tissues and, thus, not by the M. tuberculosis per se, contrary to the dictionary's definition of tuberculosis (above).

But in truth, *even the process of a microorganism's invasion into and multiplication in the body is not the cause of an infectious disease; this process is the descriptive, rather than causal, aspect of its genesis – its pathogenesis rather than etiogenesis, that is* (Section 3.5). And a distinction is to be made between infectious disease in the meaning of it being a communicable disease and it having an *infectious etiogenesis*: cervical cancer is an infectious disease not in the former but only in the latter meaning of 'infectious.'

3.7 Course

3.7.1 The Concept of Course of Illness

A case of an illness runs its course from the inception of its existence to the end of its existence – to completion of the course or to interruption of the course by death from an extraneous cause. The course of a defect-type illness, by its general essence as a state-type anomaly, tends to be quite uneventful, as in the case of various sequelae of illnesses. (But it also can be very eventful, as in the case of severe spina bifida – defective closure of the spinal column, in which a protruding sack contains the meninges, the spinal cord, or both.) In diseases and injuries, the anomaly itself, being of the process type, inherently undergoes changes over its course, but other phenomena tend to be involved besides.

As an object of medicine, inherently extrinsic to medicine itself, the course of an illness is that in the absence of medical intervention; it is the *natural course* of the illness. For this, the term in medical jargon commonly is 'natural history,' the general denotation of which is natural science in the particular meaning of "the study of animals or plants esp. as set forth for popular use" (my Concise Oxford Dictionary).

The natural course of an illness is first a matter of developments in the illness-definitional anomaly itself. Thus, in tuberculosis in the meaning of mycobacteriosis (Section 3.6.3 above), the course (natural) is prone to involve the development of new tubercles and also of fibrosis and caseation (a form of necrosis) of existing tubercles; and it also involves variation in the degree of intensity of the disease process, possibly including long periods of its 'indolence.' As for lung cancer as another example, the anomaly's natural course, as of the completion of its pathogenesis, is a matter of the malignant nodule's growth (through neoplasia), invasion of its surrounding tissues, metastasis to regional lymph nodes, and ultimately development of distant metastases, to the spine, liver, brain, and elsewhere. This natural course, if not interrupted, is ultimately fatal – hence the term 'malignant' as a descriptor of a disease like this.

While the concept of the course of an illness is, strictly speaking, the course of the illness-definitional anomaly alone, such strictness is at variance with the pragmatic outlook in medicine. The course of a hidden illness-definitional anomaly matters to the affected person – and hence also medically – not as such but only in terms of the *sickness* the anomaly itself causes and the overt manifestations also of the *complications* it causes and a possible untoward *outcome* of all of this. Sickness we've already considered, quite extensively as a matter of fact (not only in Section 3.6.3 but to an extent also in Section 3.6.3 above), but the concepts of complication and outcome of an illness have not yet been introduced.

3.7.2   Complication of Illness

My dictionaries of medicine (refs. 1 & 2 in Section 3.1) define *complication* as:

> 1. a disease or diseases concurrent with another disease. 2. the concurrence of two or more diseases in the same patient [ref. 1];

and as:

> A morbid process or event occurring during a disease that is not an essential part of the disease, although it may result from it or from independent causes [ref. 2].

My Concise Oxford Dictionary specifies the medical denotation of 'complication' as:

> a secondary disease or condition aggravating a previous one,

and my American Heritage Dictionary of the English Language specifies the medical denotation of 'complication' as:

> A condition occurring during another disease and aggravating it.

None of these definitions focuses on the concept of complication *of*, rather than in, something and, specifically, of an illness (rather than, notably, of a medical treatment). Yet, complication of an illness is a common and well-established concept in medicine. For example, The Merck Manual (1982) of mine has a section on Complications of [*sic*] Peptic Ulcer, addressing in it Penetration, Perforation, and Hemorrhage, penetration being perforation without discharge to the peritoneal sac.

But: what exactly is the concept of complication of peptic ulcer, to start with this example? Perforation of the wall, duodenal or gastric, is the most advanced stage of the ulcer's progression; and as such it is not a complication of the ulceration just as, say, extreme infirmity is but a stage, rather than a complication, of biological ageing. But peritonitis – an illness wholly distinct from peptic ulcer – when caused by the ulcer, by its perforation, clearly is a complication of the ulcer rather than a feature of the ulcer itself. Hemorrhaging from an ulcer, like hemorrhaging from an injury, is a possible feature of the anomaly itself; like perforation, it thus is not a complication of the anomaly.

As another example, tuberculosis and Kaposi's sarcoma, when caused by immunodeficiency caused by HIV infection, obviously are consequences of that infection. They are not the deficiency's manifestations; they are to the deficiency as hemorrhagic stroke is to excess in blood pressure – realizations of the state's role as a risk factor, not complications of it.

Here is my proposition, for the reader to weigh and consider: *A complication of an illness is another illness caused by it.*

3.7.3 Outcome of Illness

Outcome is left undefined in both of my medical dictionaries. My Concise Oxford Dictionary defines it as:

a result; a visible effect.

And my American Heritage Dictionary of the English Language gives, correspondingly:

A natural result; consequence.

My Oxford American Writer's Thesaurus gives more, starting with what it presents as the main synonym:

result; end result, consequence, net result, upshot, after effect, aftermath, conclusion, issue, end, end product.

For 'result' this thesaurus gives 'consequence' as the main synonym.

Consistent with (though unstated by) these definitions, I suggest for a start that outcome in general has to do with a *process*. But, different from these definitions, I suggest that it always has to do with a process in *purely descriptive* terms, with no causal implications. The outcome of a process is, I suggest, *the state in which it ends*. For example, the outcome of the human body's in-vivo course is the body's status post mortem, the lifelessness of the body representing the completion of the course of that process but not a consequence of it.

As for the course of an illness, specifically, it reaches its outcome insofar as the illness is either acute or subacute (but not genuinely chronic, as is a cancer, i.a.) and the course is not terminated by death from an extraneous cause.

There are three generic types of outcome of the course of an illness, as a matter of logical possibilities. The course (completed) may end in (the state of) full recovery of health in the sense that nothing of note is left of the illness per se (which may not be true of its complications, if any), in complete *cure*, that is. Another possibility correspondingly is (the state of) incomplete recovery, represented by some *sequela(e)* of the illness. And the third possibility is no recovery as the end state of the completed course but, instead, *fatality*/death from the illness (and hence status post mortem as the outcome state).

My dictionaries of medicine do not define recovery. *Cure* they define solely in reference to treatment. My Concise Oxford Dictionary is more explicit about this, namely that at issue is "elimination" of disease, "restoration to health," presumably by treatment.

Sequela my medical dictionaries define as:

any lesion or affection following or caused by an attack of disease [ref. 1 in Section 3.1];

and as:

A condition following as a consequence of a disease [ref. 2].

I think that the genuine concept of sequela of an illness in medicine is not that of any anomaly/illness that merely follows the one in question. For example, a case of pulmonary embolism following one of myocardial infarction is not a sequela of the MI. Nor is the concept of sequela that of an illness caused by the one in question. For example, a case of pulmonary embolism caused by deep vein thrombosis – PE occurring as a consequence of DVT – is not a sequela of the DVT; it is a complication of it (cf. above).

As I posited above, the genuine medical concept of sequela of an illness is, or at least logically would be, that which represents survival with incomplete recovery as the end state of the course (non-fatal) of an illness. For example, when recovery at the end of the process of a case of hepatitis is incomplete, this is represented by cirrhosis of the liver as the end state that it leaves behind, at the site of the process.

Only one of my two dictionaries of medicine defines *fatality*, giving the term's denotation as: "1. A condition, disease, or disaster ending in death. 2. An individual instance of death" (ref. 2 in Section 3.1). My Concise Oxford Dictionary, in turn, gives as the term's first denotation this: "an occurrence of death from an accident or in war etc." These two definitions do not upend my concept of fatality as an outcome of a case of an illness (above): death from this case of illness (and, hence, status post mortem as the outcome state of it).

3.7.4   Overview

Most broadly, then, the *course* of a case of an illness begins when the illness-definitional anomaly (somatic) comes into being (in a disease or defect at the completion of its pathogenesis, driven by its etiogenesis); and a hidden type of illness becomes consequential if and when it makes the transition from *latent* course to *overt*, sickness-manifesting course.

Apart from its resultant sickness, the course of a case of illness can be consequential in terms of the complication(s) it causes and the outcome in which it ends – fatality/death from the illness or survival with a sequela that, in its turn, is consequential in terms of sickness, complications, and/or less-than-satisfactory outcome, just as a complication of the illness itself is.

3.8   Risk

The principal object of medicine – illness – is of concern to the person affected by it, and hence to medicine, not only in such terms as it is an existing predicament to the person in question, through the suffering it is causing. The *future course* of the present illness is of great concern to a person with a case of a serious illness, and even if it still is in the latent stage of its course. And a person free of an illness can be concerned about coming down with it in the future, an overt case of it. People's concerns about the future course of their health are about such risks of possible adverse developments in it as they are aware of and find worrisome.

My dictionaries of medicine (refs. 1 & 2 in Section 3.1) define *risk* as:

a danger or hazard, the probability of suffering harm [ref. 1];

and as:

The probability that an event will occur [ref. 2].

I say, for orientation, that the concept of risk in medicine typically has to do with some untoward phenomenon (event or state) of health (illness or sickness) that a person may come to experience; it has to do with the possibility that a particular person, or any person, will come to experience it; and it is, in a particular sense, the *probability* that this experience will take place.

Risk inherently is defined in reference to a particular span of prospective time, a particular *risk period*. This may be specified in substantive terms, such as: the duration of the course of a case of an illness (as for the risk of a particular complication of it, say); the duration of remaining life (as for the risk of an overt case of a cancer, say); the duration of pregnancy (as for the risk of unwanted abortion, say); the duration of hospitalization (as for the risk of a 'nosocomial' – hospital-acquired – infection, say); etc. Alternatively, the risk period is defined quantitatively, as is that for the risk of 'neonatal death' (within 28 days from live birth), 'surgical death' (within 28 days from the operation), the 10-year risk of (an overt case of) breast cancer, etc.

An implication of the probabilistic essence of risk is that any given level of it can only be attributed to a person, or persons, with a particular *risk profile* at the beginning of the risk period, in known and relevant respects – particular realizations of certain *risk indicators*. Thus, a statement of a person's 10-year risk of breast cancer (an overt case of it) is sensible only in the context of the person's known profile in respect to the always accessible and eminently important risk indicators of gender and (attained) age; and added specificity to genetic indicators of risk (regarding the BRCA polymorphisms, most notably) and to age at first (unaborted) pregnancy, for example, is desirable, but it is not a requirement for the stated (level of) risk to make any sense.

Implicitly, any definition or perception of a level of risk is, in medicine, *predicated on not succumbing to an extraneous cause of death* within the defined risk period – which is a moot point when at issue is death from whatever cause (as in, e.g., 'neonatal death').

Risk being an individuals-oriented concept, a *population* is correspondingly characterized by the distribution of its members' thus-defined risks for the entity of health at issue, as they vary among the members with the variation of the risk profile among them based on a given set of risk indicators. And related to this, a population is characterized by the distributions of the various indicators of the risk in question.

3.9   Morbidity

3.9.1   The Prevailing Confusion

Different from an individual in isolation, a population of individuals is not mean-ingfully characterized by the presence/absence of an illness (or a sickness) at a particular moment (that of the membership-defining admission to a hospital, say); and a population's course of health over a span of time (stay in a hospital, say) is not meaningfully characterized by the occurrence/non-occurrence of a phenomenon of health (inception of an overt case of an infection, say). On the population level any phenomenon of health has the 'emergent' property of the *frequency* of its occurrence, that is, the population's level of *morbidity* from this entity.

My dictionaries of medicine (refs. 1 & 2 in Section 3.1) define morbidity as:

> a diseased condition or state, the incidence or prevalence of a disease or of all diseases in a population [ref. 1];

and as:

> 1. A diseased state. 2. The ratio of sick to well in a community. 3. The frequency of the appearance of complications following a surgical procedure or other treatment [ref. 2].

And the 'official' dictionary of epidemiology (ref. 1 in Section 3.6.1) gives under morbidity this:

> 1. Any departure, subjective or objective, from the state of physiological or psychological well-being. ...
> 2. The WHO Expert Committee on Health Statistics noted in its sixth report (1959) that morbidity could be measured in terms of three units:
>
>   (a) Persons who were ill
>   (b) The illnesses (periods or spells of illness) that these persons experienced
>   (c) The duration (days, weeks, etc.) of these illnesses

Related to item #2 in this, there is a comment about *morbidity rate*: "A term, preferably avoided, used to refer to the incidence rate and sometimes (incorrectly) to the prevalence of disease."

These definitions, and notes, individually and especially in the aggregate, are another stark reminder of the need for a thoughtful stance – rather than mere reflexive acceptance – in reading (cf. Francis Bacon; Section 2.2). As for the notion, in these notable sources, that morbidity is "a diseased condition or state," and thus "any departure ... from the state of ...," with its implicit but obvious reference to an individual's case of illness (or sickness) is not true to the term's actual usage in medicine. For, when a patient speaks of their own health, (s)he never uses the 'morbidity' term in reference to this; and likewise, when a doctor speaks about a patient's illness (or sickness), (s)he never uses the 'morbidity' term to denote this condition.

Despite the absence of such misuse of the 'morbidity' term, the term '*co-morbidity*' has recently entered into some medical usage; and it has gained entry even into one of my dictionaries of medicine (ref. 2 in Section 3.1):

> A concomitant but unrelated pathologic or disease process; usually used in epidemiology to indicate the co-existence of two or more disease processes.

It was a pioneer of 'clinical epidemiology' who adduced this odd term to denote co-illness (associated other illness, i.e.), the 'clinical epidemiology' term itself being contradiction in terms: epidemiology, I submit, is community medicine (contrasting with clinical m.; Section 4.1.3).

The WHO health statistics statement quoted above has no bearing on how morbidity – the population-level occurrence of an illness or a sickness (cf. above) – actually is quantified, and "the ratio of sick to well" is not one of the ways either. Further, morbidity is never meaningfully quantified in reference to "all diseases in a population."

3.9.2   Rates of Morbidity

Morbidity, inherently in a population, is thought of in terms of *incidence* or *prevalence* of the health entity at issue, depending on the nature of this. It is quantified in terms of a *rate* of incidence or prevalence. Like morbidity per se, rates of morbidity are a topic of considerable confusion, even among those expressly concerned with population health – epidemiologists and epidemiological researchers, that is.

The beginning of all genuine understanding of rates of morbidity is appreciation that incidence has to do with (the frequency of occurrence, in a population, of) an *event* of health, prevalence with a *state* of health. Thus, a rate of occurrence of an acute illness, by the short-duration nature of the entity at issue, is generally thought of in terms of an incidence rate, while a rate of occurrence of a chronic, long-lasting illness is commonly expressed by a prevalence rate – the rate with which it prevails in a particular population.

In the course of a case of an illness there generally is not just one type of event that can be at issue in what is said to be 'the' *incidence rate* of 'the' illness. Implied usually is that the rate, whether theoretical or empirical, quantifies the incidence of the *inception* of the illness – that of 'new' cases of it in this sense. But any empirical incidence rate actually addresses the occurrence of the *detection* of (i.e., first 'rule-in' diagnosis about) the illness. These two are very similar for illnesses that progress rapidly from the cases' inceptions to their dramatic overt manifestations, illnesses such as 'hip' (thigh-bone) fracture. On the other hand, these two rates are very different for acute illnesses that commonly do not lead to clinical detection (various communicable diseases, i.a.) and very slowly progressing chronic illnesses (prostate cancer, i.a.). Serious confusion is manifested, and caused, by writings to the effect that the introduction of a population-level program of screening for a cancer increases the incidence of the cancer when it actually increases only the rate of its detection (possibly representing, in part, pseudo-detection), mainly right after its introduction.

Similarly, in the course of a case of an illness there generally is not just one type of state (of the illness) that can be at issue in 'the' *prevalence rate* of 'the' illness. The state can be presence that is still latent but already detectable (which is relevant, notably, for decisions about screening for a cancer). Another possibility is that the rate addresses the state of the illness (e.g., tuberculosis) being in complete remission but detectable as such. Most commonly the interest is in the prevalence of a hidden illness manifest in sickness, not acute or even subacute but chronic illness or sickness (e.g., a type of arthritis or psychosis).

A rate of the occurrence of an entity of health expresses the number of these occurrences per unit amount of opportunity for the occurrence. Some rates of incidence, and all rates of prevalence, are of the form of *proportions* and, thus, dimensionless (pure numbers in the range from 0 to 1, or 0 to 100 %). Examples of these include the rate of incidence of an adverse event of health very soon after and presumably due to vaccination in a population (cohort) of vaccinated persons, and the rate of prevalence of a given type of malformation ('birth defect') among live newborns. If c cases of the entity in question occurred among N instances in which it in principle could have occurred, then the rate of the entity's occurrence is derived by carrying out the c/N division.

Other rates of incidence are derived by a division of the type of c/PT, where c again is the number of cases (of the event occurring) and PT is the amount of population-time in which the c cases of the event in question occurred. For example, if in a population of size 30 million the number of the events in one calendar year was 6,000, then the incidence rate – incidence *density* – for the event was $6,000/30(10^6 py) = 20/10^5 py$, py denoting the unit of population-time, person-year. Such a rate is not a pure number; it is dimensioned, specifically inverse-time dimensioned.

An empirical rate derived this way – that is, regardless of the population's constituent strata of age (by decade, say) and the two genders, etc. – is said to be a *crude* rate (in this sense); it is, inherently, a weighted average of the rates specific to the strata that are involved, with weights proportional to the sizes of the denominator inputs (N_j or PT_j, j = 1, 2, ...) from those strata. Deployment of some other, extrinsic weights leads to an *adjusted* rate; and two or more rates with a shared set of weights are said to be mutually *standardized* rates.

One of my dictionaries of medicine (ref. 1 in Section 3.1) defines rate as: "the number of occurrences of an event per unit time"; incidence rate it defines as "a fraction expressing the rate of new cases during a specified time period and the denominator is the population at risk during the period"; and prevalence rate it defines as "the number of people in a population who have a disease at a given time: the numerator of the rate is ... and the denominator is ... " The other one of these dictionaries presents similar ideas about rates – again to be read for weighing and considering rather than believing (Section 2.2).

3.9.3   Types of Population

The event-state duality that is at the root of the incidence-prevalence duality in rates of occurrence of entities of health is at the root, also, of a fundamental duality in the types of *population* to which these rates refer. One of these two is an *open*, or *dynamic*, population, such as the (resident) population of a city or the (citizen) population of a country. This type of population is defined by the *state* that its members represent. The population is open in the sense of being open for exit (by, e.g., death); and it is dynamic in the sense that entries together with exits imply turnover of the population's membership as it 'moves' over time.

The other type of population is a *closed* population, a *cohort*, such as a group of persons who came down with a case of some acute illness (myocardial infarction, say), an overt case of it. Membership in a cohort is clinched by an *event*, and this type of population is *closed* for exit (even by death): once a member, always a member. In consequence, all members of an MI cohort exist 1 year after the MI and are, thus, classifiable as alive or dead at that time.

My dictionaries of medicine (refs. 1 & 2 in Section 3.1) define population as: "the individuals collectively constituting a certain category or inhabiting a specified geographic area" (ref. 1); or as: "Statistical term denoting all of the objects, events, or subjects in a particular class" (ref. 2).

3.9.4   Rates vis-à-vis Risks

Rates (of morbidity) and risks (of illness/sickness) are related concepts with subtle but important differences between them.

The main relation between them is that risks – being probabilities and thus inherently only theoretical – are connected to theoretical rates and are, thus, estimated on the basis of empirical rates. The need here is to understand the nature of the *theoretical connections*, as this is a matter of elementary concepts of medicine.

The theoretical connection between risks and theoretical rates is radically different according as at issue is very short or not-very-short period of risk. In the former case the relation is very simple: the risk (probability) is equal in magnitude to the theoretical rate. For example, the risk of a given type of (sudden) adverse reaction to vaccination is the same as the theoretical (proportion-type) rate of incidence of this reaction, conditionally on the risk profile; and the risk, of ventricular fibrillation as a (sudden) complication of myocardial infarction coincides with the theoretical (proportion-type) incidence rate of this complication (conditionally on the risk profile).

When the risk period is not very short, the first subtlety in the risk is that it is to be thought of conditionally on not succumbing to death from some extraneous cause during the risk period at issue (cf. Section 3.8). And the related, added subtlety is that the risk in this context is connected to the (theoretical, survival-conditional) cumulative rate of incidence specific to particular times within the risk period.

Specifically, in this more subtle situation, the risk-determining *cumulative incidence* rate is:

$$CI_{0,T} = 1 - \exp\left(-\int_0^T ID_t \, dt\right),$$

where T is the duration of the risk period and ID_t is the (theoretical) incidence density of the event in question at time t within the risk period. It is integrated over the time span of that period.

It remains commonplace to confuse empirical rates of incidence with risks, and this is particularly misguided in the context of density-type empirical rates of incidence. Even if the theoretical incidence density of the event in question in a defined domain is some C cases per 10^5 person-years, this does not mean that the risk of the event is $C/10^5$ in reference to a one-year period of risk for persons in that domain, for whichever C.

3.10 Mortality

Related to the concept of morbidity is that of *mortality*. My dictionaries of medicine (refs. 1 & 2 in Section 3.1) define mortality as:

> 1. The quality of being mortal. 2. The mortality rate, see rate [Section 3.9.2]. 3. In life insurance, the ratio of deaths that take place to expected deaths [ref. 1];

and as:

> 1. The state of being mortal. 2. Syn mortality rate. 3. A fatal outcome [ref. 2].

I say that, in medicine, mortality, like morbidity, is a population-level concept. It is, simply, the (frequency of) *occurrence of death in a population*.

The 'official' dictionary of epidemiology (ref. 1 in Section 3.6.1) does not define mortality but it does define *mortality rate*, as:

> An estimate of the proportion of a population that dies during a specified period. The numerator is ...; the denominator is ... The death rate in a population is generally calculated by ... The rate is an estimate the person-time death rate, i.e., the death rate per 10^n person-years. ... This rate is also called the crude death rate.

But: A mortality rate actually is a measure of the level of a mortality: *it is a rate of death in the population at issue*. Like a rate of incidence of an event of illness, *a rate of death* – mortality rate – is either of the proportion type (dimensionless) or of the density type (inverse-time dimensioned). It thus is not inherently an estimate of a proportion, nor otherwise inherently an estimate. A dimensionless measure (proportion) cannot be an estimate of a dimensioned one (incidence density) and conversely. A rate does not have a numerator and a denominator: it is derived by a division, involving a numerator input and a denominator input. And a rate is not inherently a crude one. (Cf. above.)

This illustrates, further, how the concepts of morbidity – and hence of mortality too – are ones about which there is much confusion, even among epidemiologists themselves.

3.11   Glossary

In this Chapter, the existing definitions of ten key concepts of the extrinsic objects of medical thought have been critically examined and alternatives to them have been proposed. In the context of each of those concepts, a number of related concepts had to be similarly addressed.

As a review of sorts, all of those concepts are revisited here in terms of a glossary, which restates/rephrazes the proposed definitions – each of them representing my understanding of the essence of the thing at issue, of that which is true of each instance of the thing being defined and unique to it (Section 3.1). Instead of here associating any explications to these definitions, I simply refer to the Section(s) in which the concept was elaborated.

Like a dictionary in general, a glossary appended to a text is either prescriptive (normative) or merely descriptive of common usage ('structuralist'), or perhaps both. As has been amply evident from the contents of this Chapter, the remarkable truth is that, at present, *there is practically no common usage* of the English-language terms of medicine that have come up, that the terms for the most elementary concepts of medicine do not have shared meanings even for leaders of medicine whose mother tongue is English.

For this reason, a descriptive glossary of the terms that have come up is impossible to produce in a meaningful, succinct form. In this context, the definitions I put forward must be seen to be my linguistic-conceptual *propositions* for the reader to "weigh and consider" (à la Francis Bacon, Section 2.2); they are *neither descriptive nor prescriptive*.

. . .

Cause (etiogenetic, of an existing case of illness or exogenous sickness) – Antecedent (actual) such that the case would not have occurred (at the time it did) but for its presence in lieu of the defined alternative antecedent (counterfactual), ceteris paribus (Section 3.6.2).

- Direct/immediate/proximal cause – A sufficient cause or a component of it. See Indirect/mediate/distal cause (below).
- Indirect/mediate/distal cause – Cause of a cause.

- Precipitating cause – In a sufficient cause, the component closest to the effect/case.
- 'Primary cause' – Term for the inadmissible concept of the more/most important cause, falsely implying hierarchy of importance of causes. See 'Secondary cause' (below).
- 'Secondary cause' – Term for the inadmissible concept of a cause other than the 'primary' one. See 'Primary cause' (above).

Clinical (as a quality of a case of a hidden illness) – Manifest in sickness. See Overt, Sickness.

'Co-morbidity' – Misnomer for the presence, in association with a given illness, of another illness; that is, misnomer for associated illness, co-illness (Section 3.9.1). See Morbidity.

Complication (of an illness) – Another illness caused by the illness in question (Section 3.7.2).

Course (of an illness) – From the inception of a case to the end of its existence (in vivo), the pattern of changes in the anomaly itself and of the possible occurrence of complications of it, together with outcome of the course (if not interrupted by death from another cause; Section 3.7). See Complication, Outcome.

Cure (as outcome of the course of an illness) – End state of the course of an illness as (practically) complete recovery from it (Section 3.7.3). See Course, Outcome.

Determinant (of risk or morbidity/mortality) – Something on which the magnitude depends (causally or acausally).

Disease – Illness the definitional anomaly (somatic) of which is a process (Section 3.4) and the genesis of which is intrinsic to (patho)biology (Section 3.5). See Illness, Defect, Injury; Pathogenesis, Etiogenesis.

Etiogenesis/Etiology – The genesis of a somatic anomaly in (directly) causal terms; also, the counterpart of this for sickness not due to a somatic anomaly (Sections 3.5 and 3.6). See Genesis, Pathogenesis.

Etiology. See Etiogenesis.

Fatality (as a possible outcome of the course of an illness) – End of the course of an illness in death from the illness (making status post mortem the outcome state; Section 3.7.3). See Outcome.

Genesis (of an existing case of illness or of sickness not due to illness) – Coming into being, into existence (Section 3.5). See Pathogenesis, Etiogenesis.

Health – Regarding an individual, the state of (a particular part, or all, of) the soma characterized by absence of any illness (Section 3.1); also, in a broader sense, the state of the soma as to presence/absence of some particular illness (Section 3.1). Regarding a population, its level of morbidity from an illness (Section 3.9.2). See Illness, Morbidity, Soma.

Iatrogenic (as to the causation of an illness or a sickness) – Caused by a doctor's action (e.g., prescription of a medication).

Illness – Somatic anomaly which, if not overt, has at least the potential to become overtly manifest (in sickness). (Sections 3.1 and 3.2). See Sickness, Soma.

Incidence – The (frequency of) occurrence of an event (*sic*) of health in a population (Section 3.9.2). See Prevalence, Morbidity, Rate.

Incidence density – Rate of incidence (of an event of health) in reference to an aggregate of population-time (Section 3.9.2). See Population-time, Morbidity, Rate.

Latent (as a quality of a case of an illness) – Not manifest in sickness. See Overt.

Morbidity – The (frequency of) occurrence (incidence or prevalence) of an entity of health in a population (Section 3.9). See Incidence, Prevalence, Rate.

Mortality – The (frequency of) occurrence of death in a population, its incidence; that is, the death-related counterpart of morbidity from an event of health (Section 3.10). See Morbidity, Rate.

'Natural history' (of illness) – Misnomer for natural course; that is, ·for course unmodified by treatment (Section 3.7.1).

Nosocomial (as a quality of a case of an illness) – Hospital-acquired.

Outcome (of the course of an illness) – The state (or event) in which the course of an illness ends in the absence of intercurrent death (Section 3.7.3). See Cure, Sequela, Fatality.

Overt (as a quality of a hidden case of an illness or of the nature of a sign) – Manifest in sickness (re case of hidden illness) or being an element in sickness (re sign). See Latent, Sickness.

Pathogenesis – The genesis of a somatic anomaly in (purely) descriptive (acausal) terms (Section 3.5). See Genesis, Etiogenesis.

Population-time – The integral of a population's size over an interval of time (i.e., average size during the interval multiplied by the duration of the interval; Section 3.9.2).

Prevalence – The (frequency of) occurrence of a state (*sic*) of health in a population (Section 3.9.2). See Incidence, Morbidity, Rate.

Rate (of a morbidity) – A measure of the level of the morbidity in question: the number of instances of the entity (state/event) occurring divided by the amount of opportunity for its occurrence (Section 3.9.2).

- Rate of *prevalence* – The number of instances of the presence of a particular state of health in a series of person-moments (of admission to a hospital, say) divided by the size of the series; that is, of a series of person-moments, the proportion in which the state prevails.
- Rate of *incidence* – The number of instances of a particular event of health occurring in a series of fragments of person-time (hospital stay or lifetime, say) divided by the size of the series (i.e.: of a series of fragments of person-time, the proportion in which the event takes place; proportion-type incidence rate); also, the number of instances of a particular event of health occurring in an aggregate of population-time (in terms of number of person-years, say) divided by the size of that population-time (density-type incidence rate).

Risk (of a person for an adverse event or state of health) – The probability that the event/state in question will materialize, with specificity to a particular risk period (re event) or point in time (re state), and to the known risk profile of the person (Section 3.8).

Risk factor – Causal determinant of the level of risk (Section 3.6.2) or morbidity.

Risk indicator – Determinant (causal or acausal) of the level of risk or morbidity (Section 3.8).

Risk profile – The realizations of the set of risk indicators conditionally on which the level of risk is defined (Section 3.8).

Sequela – End state of the course of an illness constituted by another type of somatic anomaly (Section 3.7.3).

Sickness – The direct basis of suffering from a hidden illness, that is, the symptoms and/or overt signs manifesting their underlying (hidden) illness (Section 3.3.2); also, similar lack of well-being not representing manifestations of illness but having an extrinsic direct cause (Section 3.3.1) or a normal intrinsic one (Section.3.3.1). See Sign, Symptom.

Sign – Objective counterpart of symptom in sickness; also, abnormal result of (finding from) a diagnostic test (Sections 3.2.1 and 3.3.3). See Sickness, Symptom.

Soma (of a person) – The body (as distinct from the mind or 'soul'; Section 3.2.1).

Symptom – Subjective manifestation of (hidden) illness or of exogenous stress on healthy soma (Sections 3.2.1 and 3.3.3). See Sign.

Syndrome – Aggregate of overt/clinical features of a case which, by its nature (singular, recurrent), is conclusively indicative (pathognomic) of a singular type of its proximal cause. The cause can be congenital; otherwise a syndrome is the overt manifestation of the presence of its underlying hidden illness or an exogenous stress on healthy soma (Sections 3.3.1, 3.3.2, 3.3.3, and 3.3.4). Many so-called syndromes are not actual syndromes (Section 3.3.4). See Symptom, Sign, Illness.

Chapter 4
General Concepts of Medicine Itself

Contents

4.0 Abstract

Thought and communication in medicine presuppose general concepts of medicine and their corresponding terms, first ones having to do with things extrinsic to medicine but constituting the general objects of medical concern (Chapter 3 above); and then, obviously needed also are general concepts and terms of medicine itself.

© Springer International Publishing Switzerland 2015

O.S. Miettinen, *Medicine as a Scholarly Field: An Introduction*,
DOI 10.1007/978-3-319-19012-9_4

General concepts of medicine itself are general in the same sense as those of general objects of medicine are: they are generic and thus not specific to any particular discipline of medicine.

Like general concepts of the objects of medicine, those of medicine itself remain poorly developed. Therefore, critical examination of even the most elementary general concepts of medicine itself, and their associated terminology, is also needed in a scholarly introduction to and orientation in medicine – starting from the professional concept that the very term 'medicine' justifiably denotes. This point of departure is not only critically important but, also, quite challenging to come to grips with.

I posit a resolution to this fundamental question, and show how other concepts of medicine itself flow from this.

4.1   Medicine

4.1.1   The Essence of Medicine

My dictionaries of medicine (refs. 1 & 2 in Section 3.1) define *medicine* as:

> 1. any drug or remedy. 2. the art and science of the diagnosis and treatment of disease and the maintenance of health. 3. the treatment of disease by nonsurgical means [ref. 1];

and alternatively as:

> 1. A drug. 2. The art of preventing or curing disease; the science concerned with disease in all its relations. 3. The study and treatment of general diseases or those affecting the internal parts of the body, especially those not usually requiring surgical intervention [ref. 2].

In these two sets of definitions, our concern here is with #2 of each of them. And in weighing and considering these two conceptions of medicine at large, it is good to first focus on a single, typical discipline ('specialty') of modern medicine rather than to immediately consider medicine at large, the aggregate of its constituent disciplines.

So, do those definitions appropriately express the essence of *neurology*, to focus on this discipline of medicine as an example? As was implied in Section 3.1, the first need in this evaluation is to consider, critically, the proximate genus of this discipline of medicine, its specification in those definitions as being constituted by both art and science. For justifiably believing – or contradicting – this, a prerequisite is clarity on the respective concepts of art and science, together with clarity on whether a single discipline can be both an art and a science.

The distinction between art and science is to be made on the basis of the *respective types of product*, as has been understood ever since this distinction was introduced by the father of science (specifically biology) and of Western intellectual tradition quite generally, Aristotle. He himself studied medicine, and medicine – its theory, which he took to be the entirety of a gentleman's concern with this line of work – was taught in his Lyceum. The nature of the product of medicine – a type of ad-hoc service – makes it one of the 'productive arts,' while the product of science is abstract (divorced-from-place-and-time) knowledge. Science had nothing to do with medicine in Aristotle's time, nor with the 'humoral' ideas – pseudoknowledge – of Hippocrates that, through Galen, got to dominate European medical thought from the second century to the sixteenth, remaining very influential even later (cf. Section 1.2).

As for modern neurology, then, a distinction is to be made between this discipline of medicine on one side and neurology in the meaning of neuroscience on the other side – between neurology the art and neurology the science, that is; and this distinction is to be made with the understanding that *the science is not an aspect of the art* but extrinsic to it. The medical science of neurology serves to advance the medical art of neurology (cf. Section 1.4), to make the art more productive. But the science serving the art does not make the art a science and thereby the artist a scientist, to any extent whatsoever (cf. Section 2.4.1).

With *art* – this alone – thus established to be the proximate genus of the medical discipline of neurology and, by extension, of all genuine disciplines of medicine, it remains to weigh and consider whether it is true of all instances of the practice of a genuine art of medicine and unique to them (cf. Section 3.1) that they are about "the diagnosis and treatment of disease and the maintenance of health" or about "preventing or curing disease," or whether, instead, their essence is something else.

I suggest that something else distinguishes the art of neurology and the arts of medicine at large from those of nursing and other paramedical disciplines (and, of course, from other productive arts in general). *The always-involved, unique characteristic of the practice of an art – discipline – of medicine is, I suggest, the pivotal role, in it, of the professional's first-hand, esoteric knowing about the health of his/her client* – the pursuit and attainment of this knowing and predicating everything else in the practice on such insights into the client's health. For, any practitioner of medicine, distinct from a nurse, say, is expected to be able to teach the client (L. *doctor*, 'teacher') his/her first-hand professional insights into the health of the client – insights achieved by his/her having brought relevant *general medical knowledge* to bear on the case, on all of the facts on the client that are both available and relevant to the insights about the client's health.

This genus of first-hand, esoteric knowing in medicine – I term it *gnosis* – is esoteric in the usual meaning of this adjective: it's not directly accessible to a layperson but only to a person in possession of the requisite general knowledge, here the requisite general medical knowledge (cf. above). It encompasses not only the *diagnosis* species of it, properly construed (Section 4.2.1); its other species are *etiognosis* (Section 4.3) and *prognosis* (Section 4.4), each of these two also properly construed (Gr. *gnosis*, 'knowing,' 'knowledge,' notably when esoteric).

So, I suggest that *medicine* – in the professional, rather than pharmacological, meaning of the term – is: the aggregate of the arts of healthcare in which clients are served from the vantage of first-hand, esoteric knowing about their health. I thus make a sharp distinction between medicine and *medical science*, which I define as science to advance medicine (in that particular, professional meaning of 'medicine'). (Cf. Section 1.4.)

I thus take 'neurology' to have two, very different, denotations. For one of the entities denoted by this term the proximate genus is art/discipline (of medicine), and for the other it is science (medical). For neurology as a science, among all medical sciences, the specific difference is the demarcation of its mission: to advance the work in the art of neurology – practice in this discipline of medicine – through the advancement of knowledge (abstract) to this end. (There is, also, a third meaning of 'neurology': the science – 'pure' rather than 'applied' – concerned with the nervous system without any pragmatic purpose.)

My conception of neurology (above) is quite different from what is given in my dictionaries of medicine (refs. 1 & 2 in Section 3.1). In both of them the concept of neurology is presented as a singular one, as one of a science; but one of these (ref. 1) adds, quite incongruously with the overall definition, that "clinical" neurology is "that specialty concerned with the diagnosis and treatment of disorders of the nervous system."

The concept of cardiology, too, both of those dictionaries present as being a singular one, but they are strikingly discordant about the proximate genus of this field: according to one of them (ref. 1), cardiology is a "study," while per the other it is a "medical specialty." Pulmonology – the other one of the two preeminent disciplines of medicine concerned with the chest – the former source again presents as a singular one, but its proximate genus it gives as "science" (in contrast to "study" for cardiology), while the other dictionary does not have the 'pulmonology' entry!

Let's consider, also, the principal discipline of medicine having to do with the abdominal region of the body: Gastroenterology is said to be "the study [*sic*] of the stomach and intestines and their diseases" (ref. 1), or "The medical specialty [*sic*] concerned with the function and disorders of the gastrointestinal tract, including stomach, intestines, and associated organs" (ref. 2).

One more: the mother of cardiology, pulmonology, gastroenterology, etc. – internal medicine. It is said to be a "medical specialty" by one of those dictionaries (ref. 1) and a "branch of medicine" by the other (ref. 2); it thus is not a study nor a science according to either one of these. The "medical specialty" called internal medicine is said to be one "dealing especially with the diagnosis and medical treatment of diseases and disorders of the internal structures of the human body," while the "branch of medicine" called internal medicine is said to be "concerned with nonsurgical diseases in adults, but not including diseases limited to the skin or to the nervous system." The concept of branch of medicine and those of nonsurgical and surgical diseases this latter dictionary (ref. 2) leaves undefined.

Whereas I define medicine quite differently from the ways it is defined in my dictionaries of medicine (above), my definitions of the disciplines that jointly constitute medicine are also different: In each of them the proximate genus naturally is specified as discipline ('specialty') of medicine, and *the difference specific to any given one of the disciplines has to do with the gnostic concerns in it.* Appropriate definitions/demarcations of those discipline-specific domains of gnosis I address in Section 7.2.

In the framework of my definition of medicine (above), and taking the concept of physician to be the conventional one (i.e., any graduate of medical school licensed to practice medicine), a *physician's practice of 'medicine' may not be practice of genuine medicine*. For example, physicians practicing diagnostic radiology do not practice the pursuit of diagnosis but merely provide for actual diagnosticians – 'bedside' doctors, say – relevant facts as inputs to their diagnoses, just as doctors' clients and the personnel in non-radiological diagnostic laboratories do.

4.1.2   Surgery in Medicine

As for *surgery*, then, my dictionaries of medicine define it as:

> that branch of medicine which treats diseases, injuries, and deformities by manual or operative methods [ref. 1 in Section 3.1];

and as:

> The branch of medicine concerned with the treatment of disease, injury, and deformity by operation or manipulation [ref. 2].

Thus, according to both of these dictionaries, surgery is intrinsic to medicine rather than lateral to it (à la 'medicine and surgery'). As a notable aside here, both of these definitions of surgery present the concern in this "branch of medicine" as being with illness at large rather than disease alone (cf. Sections 3.2, 3.3, and 3.4).

I take exception to those definitions of surgery, to their presentation of surgical treatment as being in the essence of a "branch" of medicine rather than, merely, a particular *modality of treatment* of concern in (the practice of) medicine, along with medicational treatment, radiological treatment, and others; for, medicational treatment, for example, is not taken to be definitional to another branch of medicine, nor should it be.

In *genuine disciplines/branches of medicine* that have anything to do with treatment, all of the available options for treatment are considered (under prognosis) – some of the options can be surgical, some medicational, and yet others neither; and some of them can be combinations of such modalities – and a choice is made among them, perhaps by the patient/client alone, upon the doctor's 'doctoring' (teaching the patient, that is; Section 2.1). Upon this doctoring, the doctor may carry out the chosen treatment, rather than have someone else – commonly the patient – carry it out.

So, as I argued above, it is the pivotal role of gnosis that is in the essence of (whichever genuine discipline of) medicine, not any action predicated on it, including action taken by the doctor him- or herself. When the doctor, predicated on his/her own, first-hand prognosis carries out a treatment, this I take to be an incidental element in his/her practice of medicine and thus not definitional to – in the essence of – medicine. The treatment may be a surgical operation, the injection/infusion of a medication, or whatever.

In the context of a concept, logicians distinguish between the inherent *properties* of the thing in question and *accidentals* that may in some instances characterize it (ref. 3 in Section 3.1). The thing's properties flow from – can be deduced from – the essence of it. From the gnostic essence of medicine (above) flows, for example, the property of medicine that the doctor, in certain situations, necessarily addresses the prognostic implications of the choice of treatment; but the doctor's execution of the chosen treatment is but an accidental in medicine.

4.1.3 Branches of Medicine

Rather than by modality of treatment, I distinguish between the branches of medicine at large – regions of the field in which particular disciplines of medicine fall – according to a *fundamental duality in the types of doctors' clients*. I thus take the principal branches of medicine to be clinical medicine and community medicine.

Clinical medicine my dictionaries of medicine (refs. 1 & 2 in Section 3.1) define as:

the study of disease by direct observation of the living patient [ref. 1];

and as:

the study and practice of m. [medicine] in relation to the care of patients; the art of m. as distinguished from laboratory science [ref. 2].

But to me, clinical medicine is medicine (proximate genus) in which the *clients are individuals*, served as such, one at a time (specific difference; Section 3.1).

Community medicine is defined in only one of those two dictionaries, as:

the study of health and disease in a defined community; the practice of m. in such a setting [ref. 2].

But my conception of this, too, is quite different from that in this "authoritative" dictionary of medicine. To me, community medicine is medicine in which *the client is singular: the population of a community*, served as such (rather than serving its members individually).

To me, 'community medicine' is synonymous with '*epidemiology*.' But my dictionaries of medicine define epidemiology as:

the science concerned with the study of the factors determining and influencing the frequency and distribution of disease, injury, and other health-related events and their causes in a defined human population for the purpose of establishing programs to prevent and control their development and spread. Also, the sum of knowledge gained in such a study [ref. 1 in Section 3.1];

and alternatively as:

The study of the distribution and determinants of health-related states or events in specified populations, and the application of this study to control of health problems [ref. 2].

The prevailing confusion about the essence of epidemiology is addressed at length in the opening chapter of a textbook on epidemiological research (Miettinen OS, Karp I. *Epidemiological Research: An Introduction*. New York: Springer, 2012). It does not merit elaboration here.

4.1.4 Rational Medicine

The supremely important concept of *rational medicine* (cf. Section 1.1) is defined in only one of those dictionaries of medicine, thus:

> practice of medicine based upon actual knowledge; opposed to empiricism [ref. 1 in Section 3.1];

and *empiricism* it defines as:

> 1. The method [*sic*] of the empiric school of medicine; opposed to rational medicine. 2. Reliance on mere experience; empirical practice. 3. Quackery.

This I view as a distorted (and disparaging) echo of the essence of the Hippocratic school of medicine – for which commitment to rationality in learning from experience is definitional (Section 1.1). And this misrepresents also the concept of empiricism in the philosophy of science, where it has to do with the source of knowledge but has nothing to do with the knowledge not being "actual" (ref. 1 below):

> Empiricism is the belief that all our knowledge derives from the experience of our senses (or from extensions of our senses in instruments). ... Philosophers have seen empiricism as being opposed to rationalism, the belief that reason alone can arrive at basic truths about the world, ... Kant attempted a reconciliation of rationalism and empiricism. ...

Reference 1: Laudan R. Empiricism. *In*: Heilbron JL (Editor). *The Oxford Companion to the History of Modern Science*. Oxford: Oxford University Press, 2003.

For rational medicine, I say, the true alternative is not empiricism but *irrational medicine*, or illogical medicine. I note that my Oxford American Writer's Thesaurus gives 'logical' as a synonym of 'rational,' and 'illogical' as its antonym. Under 'Sensible' it explains that "rational suggests the ability to reason logically and to draw conclusions from inferences (a rational explanation for ...)." The central role of observation – this empirical input to learning in conjunction with rational thinking – has not made Hippocratic medicine (Section 1.1) irrational and, thereby, tantamount to quackery (cf. above). Irrational would be, instead, the idea that "actual knowledge" of medicine is produced only by medical sciences and that it was, therefore, nonexistent before the advent of these sciences (2.4.1; cf. Sections 1.3 and 1.4).

4.1.5   Scientific Medicine

The concept of rational medicine can best be understood, I suggest, in the context of a tenable conception of *scientific medicine*, which neither one of my dictionaries of medicine defines. While this term has been used in many ways, two conceptions of scientific medicine came to eminence in the twentieth century, both remaining very influential at present. I, however, take exception to both of these; and a singular, tenable conception of scientific medicine I take to be essential for modern thought about medicine as a scholarly field.

One of those two conceptions of scientific medicine arose from the concern in the American Medical Association that there were, at the turn of the twentieth century, too many 'unscientific,' quackery-cultivating medical schools in the U.S. So the AMA instigated the development of the *'Flexner report,'* published in 1910 (ref. 2 below). According to Flexner, scientific medicine is characterized by scientific-type thinking in its problem-solving, which medical students can learn by following the work of [whichever] art scientists in the laboratories of medical research. Thus, Flexnerian medicine could be termed *thinking-based medicine.* Today's medical education, however, involves study of the knowledge produced by research in the 'basic' medical sciences without any attention to the patterns of thought in the research behind that knowledge (Section 2.4.1; cf. 'actual knowledge' in 'rational medicine' in Section 4.1.4).

Reference 2: Flexner A. *Medical Education in the United States and Canada.* Bulleting no. 4. New York: Carnegie Foundation for the Advancement of Teaching, 1910.

The other one of those two conceptions of scientific medicine was adduced in 1992 (ref. 3 below):

> [The old] paradigm puts a high value on traditional scientific authority and adherence to standard approaches, and answers are frequently sought from direct contact with local experts or references to writings of international experts. The new paradigm puts a much lower value on authority. The underlying belief is that physicians can gain the skills to make independent assessments of evidence and thus evaluate the credibility of opinions being offered by experts. It follows that clinicians should regularly consult the original literature ... in solving clinical problems and providing optimal patent care.

Reference 3: Evidence-Based Medicine Working Group. Evidence-based medicine: a new approach to teaching the practice of medicine. *J Amer Med Assoc* 1992; 268: 2420–2425.

The *evidence-based medicine* movement that thus was launched has grown to the point at which future practitioners of medicine are commonly expected to study textbooks of 'clinical epidemiology' and of 'evidence-based medicine' in order to "gain the skills to make independent assessments of the evidence ..." etc. (cf. above). But preparatory to critically weighing and considering these expectations, it might be good to study my book *Up from Clinical Epidemiology & EBM* (New York: Springer, 2011), which recounts my reflections on all of this.

My conception of scientific medicine is, as I remarked above, very different from both of those two. In my conception, scientific medicine is medicine with this specific difference distinguishing it from non-scientific medicine: It has a *rational theoretical framework* and, in this framework, a *knowledge-base from science* – from what I term 'quintessentially applied medical research.' (Cf. Section 1.4.) Only in this conception of it is scientific medicine *knowledge-based medicine*. (The etymology of 'science' has to do with knowledge.)

In these terms, today's medicine still is, in all essence, nonscientific. But the path to scientific medicine, scientific in this sense, has been charted in my *Toward Scientific Medicine* (New York: Springer, 2014). The implications of this vision are beyond the concerns of a beginning student of medicine, except if (s)he actually is entertaining the possibility of a career in quintessentially applied medical research instead of medicine itself.

4.1.6   Pseudomedicine

Given the central role of general medical knowledge in the essence of medicine (Section 4.1.1), *genuine medicine is knowledge-based medicine*, and in a particular meaning of this: A doctor in the practice of genuine medicine brings general medical knowledge to bear on the available facts on the case at hand so as to achieve *esoteric knowing* – gnosis – about the health of the client beyond those facts; and (s)he then 'doctors' (teaches) the client accordingly.

From this conception of genuine medicine flows the corresponding concept of *pseudomedicine*: affectation of the practice of genuine medicine, specifically affectation of possession of the existing, relevant general medical knowledge, while actually being devoid of this.

Pseudomedicine, in this sense of it, is generally termed *quackery*. My dictionaries of medicine (refs. 1 & 2 in Section 3.1) define quackery as:

> the fraudulent misrepresentation of one's ability and experience in the diagnosis and treatment of disease or of the effects to be achieved by the treatment offered [ref. 1];

and as *charlatanism*:

> a fraudulent claim to medical knowledge; treating the sick without knowledge of medicine or authority to practice medicine [ref. 2].

These definitions of pseudomedicine are less than fully consistent with their corresponding definitions of genuine medicine in Section 4.1.1; but one of the inconsistencies is very positive: that second definition of quackery (from Stedman's) implies that deployment of *knowledge* is in the essence of genuine medicine (cf. Section 4.1.1). Neither one of the corresponding direct definitions of medicine expresses deployment of knowledge as an essential feature of (genuine) medicine.

That same definition of quackery/charlatanism, however, introduces a *highly questionable idea* about this and, thereby, about the essence of medicine proper: It implies that *unauthorized* practice of (even) knowledge-based medicine is quackery, while authorized practice of (even) knowledge-blind medicine is genuine medicine – that is, that medicine is whatever authorized practitioners of medicine do under the guise of 'medicine,' while whatever others do is not medicine.

For a beginning student of medicine there is no topic of needed learning more critically important than *the essence of medicine*, of genuine medicine, that is; and this (s)he cannot learn from "authoritative" dictionaries of medicine, not directly (Section 4.1.1) nor indirectly under 'quackery.' But having learned the essence of genuine medicine, (s)he needs to understand that medicine is what it is (by a tenable definition; Section 4.1.1) regardless of whether it is authorized or unauthorized, well-informed or ill-informed, good or bad. (Cf. concept of doctor.) And (s)he needs to understand that there are many species of pseudomedicine other than quackery, generally representing misunderstanding rather than fraud – 'experimental medicine,' 'nuclear medicine,' etc.

4.2   Diagnosis

4.2.1   The Essence of Diagnosis

In medicine, the concept of diagnosis is central to the point of being entered into the very definition of this aggregate of arts in one of my two dictionaries of medicine (Section 4.1.1). Those dictionaries (refs. 1 & 2 in Section 3.1) define *diagnosis* as:

> 1. the determination of the nature of a case of disease. 2. the art of distinguishing one disease from another [ref. 1];

and as:

> The determination of the nature of a disease [ref. 2].

So, by these definitions the proximate genus of diagnosis is the process entity determination; and the specific difference of diagnosis within this proximate genus those dictionaries of medicine naturally present as that which they take diagnosis to determine, namely the nature of a case of illness (disease or other; cf. Section 3.4). Implied by this is that diagnosis inherently has to do with situations in which some illness is known to be present.

That conception of the essence of diagnosis a student of medicine needs to weigh and consider in the context of what I said, in Section 4.1.1, about the essence of medicine, namely that: "The always-involved, unique characteristic of [whichever] art ... of medicine is ... the pivotal role, in it, of the professional's *first-hand, esoteric knowing* about the health of his/her client." One species of such knowing is diagnosis (Section 4.1.1).

Regardless of whether the proximate genus of diagnosis is seen to be determination or knowing, it is definitely untrue that diagnosis always is about the nature of the illness whose presence is a given. Even when the prompting for the diagnostic pursuit is the client's complaint about a sickness, there may be no illness at the root of it: the sickness can be environmental or behavioral rather than constitutional/somatic in its direct causation (Section 3.3.1); or the complaint possibly is a case of mere malingering. Moreover, the presence of some illness is not a premise for diagnosis in the absence of any relevant complaint – in the pursuit of various 'rule-out' diagnoses for insurance or employment, say, or of 'rule-in' diagnosis of a particular illness in screening for the illness.

So, I suggest that diagnosis is not, by any reasonable general definition, about the nature of the illness on the premise that some illness is present. Instead, I suggest, it generally is about the *presence/absence of a particular illness*, expressly targeted for diagnosis. The result of the diagnostic pursuit can be practically conclusive, a practical 'rule-in' or 'rule-out' diagnosis about the illness targeted for diagnosis; that is, the result can be diagnosis about the illness in question with probability about its presence very high or very low (or, correspondingly, about its absence very low or very high).

And in general, thus, I take diagnosis to be a doctor's first-hand, esoteric *knowing about the probability* of the presence (or absence) of a particular illness. Implied by this is that there in principle is the correct probability that the knowing is about (Section 5.2).

4.2.2 Differential Diagnosis

When the premise actually is that some illness (hidden) is present (as the explanation for the patient's sickness) and the diagnosis consequently is to be about the nature of that illness, the point of departure naturally is a conception of what the various possibilities are; and the diagnosis is about each of these possibilities, generally with the idea that the higher is the probability for the presence of a particular one of these illnesses, the lower it consequently is for the others being present. But the knowing in diagnosis needs to be about each of the possibilities.

Diagnosis in this situation, very common, inherently is differential among the possibilities in the *differential-diagnostic set* of these, and the diagnosis about any one of the illnesses in this set can thus be termed differential diagnosis.

Even though both of my dictionaries of medicine define diagnosis as though it inherently were differential diagnosis, both of them also give a definition of *differential diagnosis* separately from that of diagnosis. One of them is this:

> the determination of which of two or more diseases with similar symptoms [i.e., in the differential-diagnostic set] is the one from which the patient is suffering, by a systematic comparison and contrasting of the clinical findings [ref. 2 in Section 3.1].

The other definition is almost the same as this.

Particularly notable about this definition of diagnosis (inherently differential in the context of determining the nature of the patient's illness) is that it specifies (vaguely) the means or *method* of the "determination": "a systematic comparison and contrasting of the clinical findings." But, with diagnosis construed as a species of esoteric knowing, express *principles* of thinking about the knowledge-base of it are the counterpart of this (nebulous) process of "comparison and contrasting." (Section 5.2.)

4.2.3   Clinical Diagnosis

The essence of diagnosis as a species of knowing comes to focus by considering, specifically, *clinical diagnosis*. My dictionaries of medicine (refs. 1 & 2 in Section 3.1) define it as:

> diagnosis based on signs, symptoms, and laboratory findings during life [ref. 1];

and as:

> a [diagnosis] made from a study of the signs and symptoms of a disease [ref. 2].

The concepts of symptom and sign were addressed under Sickness in Section 3.3.2, as ones that, while not intrinsic to medicine, are of great concern in medicine, specifically in clinical medicine (defined in Section 4.1.3).

In weighing and considering those definitions of clinical diagnosis, the latter one of them specifically, the key question is whether clinical diagnosis truly is "based on" or "made from" the *diagnostic profile* constituted by the available clinical facts on the case, assembled in the processes of history-taking, physical examination, and clinical (rather than laboratory) testing. The truth is that genuine diagnosis is a species of knowing and that it is not "made from" this set of facts: it is *based on* these facts, but in part only. The added, crucial basis for the esoteric knowing that clinical diagnosis represents is *general medical knowledge* (Section 4.1.1) – about the frequency/*probability* with which the illness in question is present in instances of the diagnostic profile in general (in the abstract).

Diagnosis with inputs from laboratory tests (in addition to the clinical facts) is devoid of an established term to denote it. It could be termed *meta-clinical* diagnosis.

The theory of diagnosis I address further in Sections 5.2 and 8.1 and in Appendix 1.

4.2.4   Pattern Recognition

When the illness whose present/absence is the object of diagnosis is a *hidden anomaly* of the client's soma (Section 3.2.2), knowing about the truth in question is aptly termed *dia-gnosis* – 'through-knowing' – as the truth at issue is not immanent in the set of facts constituting the diagnostic profile but is deeper than this: it is only perceived, probabilistically, through that profile, in the light (dimly) of general medical knowledge (Sections 4.2.1 and 4.2.3 above). Diagnoses in 'internal medicine' and in the various subdisciplines of this are, by definition, diagnoses about hidden anomalies (somatic, illness-definitional).

What, then, about *overt* illness-definitional anomalies (somatic) such as those that typically are of concern in dermatology? It is commonplace to say that the doctor 'diagnoses' psoriasis, for example; but in point of fact (s)he merely recognizes the pattern of the lesion(s) to be that of so-called psoriasis and labels the illness accordingly. This is *pattern recognition, not diagnosis* in the meaning above (and in Sections 4.2.1 and 4.2.3 above). The pattern of this illness is not thought of as being pathognomonic (conclusively indicative) of the presence of a deeper anomaly of concern – very different from, say, Cushing's syndrome as being diagnostic about the presence of a certain type of endocrine anomaly (when not due to medication use).

Akin to this, a psychiatrist is said to 'diagnose' a case of depression, for example; but this too is a matter of labelling the set of facts on the surface, of the patient's sickness, rather than 'through-knowing' about a deeper, illness-definitional truth achieved by bringing general medical knowledge to bear on the diagnostic profile of the case.

And a radiologist or a pathologist may 'diagnose' a particular malignancy, say, on the basis of diagnostic testing (imaging or biopsy) alone; but at issue is merely a pattern of diagnosis-relevant facts. However strongly indicative of malignancy the pattern is, this in itself is not malignant: true diagnosis of malignancy is inferential beyond whatever facts in the diagnostic profile.

All in all, thus, a distinction is to be made between actual diagnosis and *so-called diagnosis constituted by mere pattern-recognition* in the available facts.

4.3 Etiognosis

Given a practical 'rule-in' diagnosis about a particular illness, knowing about some component of the sufficient – direct/immediate/proximal – cause of the illness (Section 3.6.1) in the case at issue may be seen to be relevant for further dealing with the case; and concern to know about the proximal sufficient cause is a concern in general when it has to do with sickness presumed not to be manifesting an underlying illness but being caused by some circumstantial stressor (environmental and/or behavioral, impinging on healthy soma; Section 3.3.1).

Knowing about the causal origin – etiology/etiogenesis (Section 3.6.1) – of a case of illness, or of sickness without illness (Section 3.3.1), is naturally termed *etiognosis*, where the 'gnosis' element denotes esoteric knowing as in 'diagnosis' (Section 4.2.1) and the prefix refers to causation, as in 'etiology' and 'etiogenesis' (Section 3.6.1). Despite the usefulness of the concept, the 'etiognosis' term cannot be found in my dictionaries of medicine (refs. 1 & 2 in Section 3.1), nor is the concept known by any other term. This 'etiognosis' term is a rather recent neologism I have adduced.

The knowing in etiognosis, akin to that in diagnosis, has as its context the profile – here the *etiognostic profile* – of the case, constituted by the species of illness/sickness, the potentially etiogenetic history of the patient, and other relevant features of it. Etiognosis conditional on this profile is a matter of esoteric knowing – probabilistic, as is diagnosis – about whether a particular, potentially causal antecedent, present in lieu of a particular alternative of it (Section 3.6.1), actually was causal to the case – this as a matter of *general medical knowledge* specific to an antecedent that was present before the case at issue.

More on etiognosis in Section 5.3.

4.4 Prognosis

It may be recalled from Section 4.1.1 that diagnosis is involved in one of the definitions of medicine in my two dictionaries of medicine, and the point of note now is that prognosis is not involved in either one of them. The reason for this can be seen to be that according to those dictionaries – and indeed in medicine quite generally (though unjustifiably) – diagnosis is thought of as the "determination" of the nature of the patient's present illness, while nobody thinks of prognosis as the determination the patient's future health. Moreover, while diagnosis is commonly ascribed to the doctor, it is commonplace to think of prognosis as a property of the case at issue rather than anything having to do with the doctor's knowing about it, as when saying that a particular illness (*sic*) has a good, or a bad, prognosis.

Those two dictionaries of medicine define *prognosis* as:

> a forecast as to the probable outcome of an attack of disease; the prospect as to recovery from a disease as indicated by the nature and symptoms of the case [ref. 1 in Section 3.1];

and as:

> A forecast of the probable course and/or outcome of a disease [ref. 2 there].

So in these definitions the proximate genus of prognosis is the doctor's forecast (about the future), different from determination (about the present) in the context of diagnosis.

My Concise Oxford Dictionary defines *forecast* as: "calculation or estimate of something future, esp. coming weather;" and *probable* it specifies as denoting the quality of something such that it "may be expected to happen or proven true; likely." Forecast specifying something to be the probable outcome is a *prediction* of that outcome, foretelling that outcome – founded on the presumption of having foreknowledge of what the outcome will be. So, just as my medical dictionaries define diagnosis in terms that imply attainment of quite categorical knowing about the present, they define prognosis in terms that imply equally categorical knowing about the future. But this is not realistic.

My conception of prognosis is quite analogous to its counterpart in respect to diagnosis. Prognosis as a concept in medicine I take to be *knowing* about the future occurrence of a particular entity of health, first-hand esoteric knowing about the *probability* of its occurrence – whatever the level of this probability may be.

The conditionality of prognosis is, however, not wholly analogous to that of diagnosis, as matters of *causality* – effects of intervention – are commonly involved in modern prognoses. This was addressed, to some extent, under Risk (Section 3.8).

More on prognosis in Section 5.4.

4.5   Knowledge

4.5.1   *'Actual Knowledge' in Medicine*

While pivotal in the practice of medicine I take to be the pursuit and attainment of esoteric knowing about the client's health, with everything else in doctors' practices predicated on this essential element (Section 4.1.1), and while this esoteric ad-hoc knowing – gnosis – requires general medical knowledge about the probabilities at issue (Sections 4.2, 4.3, and 4.4 above), the question arises: What is the nature of this general medical knowledge? and specifically, *By what criterion is a particular gnostic probability a matter of general medical knowledge?*

Even though I thus view medicine as an aggregate of centrally knowledge-dependent professions, *the status accorded to gnosis-serving general knowledge in medicine remains remarkably ambiguous.* My dictionaries of medicine (refs. 1 & 2 in Section 3.1) do not define knowledge; in neither one of those two dictionaries is diagnosis defined in terms of knowledge (Section 4.2.1), and the same is true of prognosis (Section 4.4 above); and knowledge is not involved, even, in either one of the two conceptions of scientific medicine that came to eminence in the twentieth century (Section 4.1.5).

While knowledge thus seems to be given no role in medicine, one of my dictionaries of medicine defines rational medicine as being characterized by the deployment of "actual knowledge," contrasting medicine based on actual knowledge with quackery (Section 4.1.4). But in this, the meaning of actual knowledge as the basis of medicine appears to be knowledge from the 'basic' sciences of medicine (Section 2.4.1) rather than knowledge pertaining to gnostic probabilities, notably as neither diagnosis nor prognosis is commonly defined in probabilistic terms. Yet, I firmly hold what I say about the knowledge-based – gnosis-related – essence of medicine (Section 4.1.1).

If, in the context of a given case profile, *experts* on that type of case quite generally set substantially discordant values for the gnostic probability at issue (diagnostic, say), it seems fair to say that, for a case like this, the requisite general medical knowledge does not exist (instead of saying that the relevant knowledge from 'basic' medical sciences does exist but its gnostic implications remain indefinite). And by the same token, existence of the requisite medical knowledge for a given gnosis (profile-conditional) would reasonably be seen to be defined by *experts' shared belief* about the level of the probability in question, shared to some reasonable degree of convergence of the individual beliefs.

In these terms, the 'actual knowledge' of *genuinely rational medicine* includes knowledge about the nonexistence of the knowledge relevant to the gnosis at issue in given types of case; and practice of rational medicine in those situations involves not only knowing the nonexistence of the relevant knowledge but also acting accordingly, starting with acknowledgment of this in the 'doctoring'/teaching of the client. Such practice indeed contrasts with the pretenses characteristic of quackery (cf. above).

When knowledge in the sense of experts' shared belief does not exist, *ersatz knowledge* can take the form of experts' typical belief. But an individual doctor's personal belief about the level of the probability in question is an *opinion*, not knowledge, even if commonly thought of as 'tacit knowledge.' And gnostic probabilities in "evidence-based medicine" (Section 4.1.5) also are opinions (personal, informed by evidence), not knowledge.

4.5.2   Hierarchy of the Knowledge

Confronted with a given case in the practice of medicine, the *principal knowledge-related – profile-conditional – challenge* to the doctor thus is threefold: need to know what probabilities (gnostic) should be known; need to know whether knowledge about these probabilities exists; and insofar as the knowledge does exist, need to know what the probabilities – experts' shared beliefs (Section 4.5.1 above) about their levels – actually are.

As these elements of practice-relevant knowledge are specific to the gnostic profile of the case at issue, their antecedent, first-order need is to know *what facts* are to be assembled for the profile of the case.

4.6   Treatment

4.6.1   Species of Treatment

In the definitions of medicine in my dictionaries of medicine (different from what I propose; Section 4.1.1), considerable eminence is given to treatment (even though those definitions make no reference to prognosis, for which knowledge about treatment effects is needed; Section 4.4).

Those dictionaries define *treatment* as:

the management and care of a patient for the purpose of combating disease or disorder [ref. 1 in Section 3.1];

and as:

Medical or surgical management of a patient [ref. 2].

Management is left undefined in both of those dictionaries, while care is defined in one of them as "the services rendered by members of the health professions for the benefit of the patient" (ref. 1), and as "a general term for the application of knowledge [*sic*; cf. Section 4.5.1] to the benefit of a community or individual" (ref. 2).

For the related concept of *therapy* the respective definitions are:

the treatment of disease [ref. 1]

and

The treatment of disease or disorder [ref. 2].

And *intervention* is defined differently from both treatment and therapy, as:

1. the act or fact of interfering so as to modify. 2. specifically, any measure whose purpose is to improve health or to alter the course of disease [ref. 1]

and as

an action or ministration that produces an effect or that is intended to alter the course of a pathologic process [ref. 2].

This set of definitions of these three concepts – each of them very central in medicine – a thoughtful student of medicine is prone to find quite confusing. Is 'therapy' indeed, as implied here, a synonym of 'treatment' and 'management'? And what really is intervention in relation to these concepts?

I suggest a way out of this confusion: The genus of medical action at issue here is *treatment*; and its species are naturally defined on the basis of the *purpose* of it, according as the purpose is to ameliorate the patient's existing sickness, or to improve the course of the patient's existing illness, or to reduce the client's risk for future illness. Treatment directed to existing sickness is *palliative* in its purpose; treatment directed to existing illness is *therapeutic* in its purpose; and treatment aimed at reduction of the risk of future illness or sickness not due to illness is *preventive/prophylactic* in its purpose. *Intervention* is treatment directed to the course of health, and it therefore is either prophylactic or therapeutic (rather than merely palliative).

Treatment is an element – possible rather than routine (Section 4.1.1) – in medicine, in the overall service by a doctor to a client. This service is not 'management' of the client, nor of the client's health; for in principle at least, the recipient of the service is in charge, not the provider of it (Sections 6.2 and 6.3).

It may be noted that treatment (medical) is something that is done to a person's body directly, and that treatment (of the body/soma/constitution) typically is medicational, while health-oriented change of environment (as to climate, say) or behavior (dietary, say) – typically preventive – is not treatment.

4.6.2   Rational Treatment

Rational treatment – echoing the topic of "rational medicine" (Sections 4.1.4 and 4.5.1) – is brought up under Treatment in my Dorland's (ref. 1 in Section 3.1) and under Therapy in my Stedman's (ref. 2 in Section 3.1). The former source defines it as:

> treatment based upon a knowledge of disease and the action of the remedies employed,

which contrasts quite starkly with what is given in the latter source:

> therapeutic procedures introduced by Albert Ellis and based on the premise that lack of information or illogical thought are basic causes of a patient's difficulties; it is assumed that the patient can be assisted in overcoming his or her problems by a direct, prescriptive, advise-giving approach by a therapist.

I suggest that treatment (interventive) is rational only if the decision about it is taken in a rational theoretical framework for prognosis together with appreciation, in this framework, of the state of the knowledge-base for prognosis in the type of case (in the type of setting of the care) and with due appreciation of the relevant valuations and preferences of the client. (Cf. scientific medicine in Section 4.1.5 and prognosis in Sections 4.4 and 5.4.)

Related to this, I suggest that treatment is *pseudorational* if it is chosen in the decision-theoretical framework advocated by health economists (Section 5.8.3).

4.7   Glossary

In Section 3.11, the glossary of terms and concepts of medical thought about matters extrinsic to medicine itself was preceded by a preamble. That preamble applies, mutatis mutandis (i.e., upon making the necessary alterations), also to terms and concepts intrinsic to medicine, ones having to do with medicine itself.

Client (in medicine) – A person (in clinical medicine) or a population (in community medicine) served by a doctor. See Doctor, Patient.

Diagnosis (in clinical medicine) – Gnosis about the presence/absence (current or past) of a particular illness in a client. (Section 4.2.1.) See Gnosis, Client, Patient.

Diagnostic/pathognomonic (as a quality of an element in, or the entirety of, a diagnostic profile) – See Pathognomic. Note: In another meaning of the term, a diagnostic is a procedure to produce supplementary information into the diagnostic profile (e.g., stress test or angiography, to supplement clinical fact-finding, for a diagnosis about coronary stenosis).

Doctor (in clinical medicine) – For a recipient of clinicians' care, the one who knows, at least as well as anyone else involved, about his/her health (in a particular respect) and is, on this basis, his/her teacher about it. See Client, Patient.

Esoteric: See Gnosis.

Etiognosis (in clinical medicine) – Gnosis about the etiogenesis of a case of illness (disease or defect) or exogenous sickness (not due to illness), specifically about whether a particular antecedent/concomitant of the case was/is causal to it. (Section 4.3.) See Gnosis.

Forecast/prediction (in general) – Pronouncement/belief that a particular event/state will take place. (Section 4.4.) See Prognosis.

Gnosis (in clinical medicine) – Knowing (first-hand, esoteric) about the health of a client beyond the available facts on him/her, on the basis of general medical knowledge applied to these. Possession of the requisite general medical knowledge is unique to doctors (in the relevant discipline of medicine); that knowledge is *esoteric* in this sense. (Sections 4.1.1 and 5.5) See Client, Patient; Knowing, Knowledge; Diagnosis, Etiognosis, Prognosis.

Intervention (in clinical medicine) – Action on a client's soma considered/adopted as a means to change the course of the client's health for the better. (Section 4.6.1.) See Client, Patient. Notes:

- Change of a client's/patient's environment or lifestyle is not a doctor's action, nor anyone's action, on their soma; it is not an intervention (on the person's course of health; Section 4.6.1).
- Intervention (in clinical medicine) – typically medicational – is but rarely executed by the doctor who teaches about it. (It usually is executed by the patient/client).

Knowing – Possession of knowledge (particularistic or general/abstract). See Knowledge, Gnosis.

Knowledge (general/abstract, in medicine) – Experts' typical, essentially convergent belief about the truth in question. (Section 4.5.1.) See Knowing, Gnosis.

Medicine – The aggregate of the professional arts of healthcare in the essence of which is the professional's – doctor's – pursuit and attainment of first-hand esoteric knowing (gnosis) about the health of the client beyond the available facts on the case; also, the practice of a given one of these arts. (Section 4.1.1.)

- *Clinical* m. – Medicine in which the doctor serves individuals, one at a time. (Section 4.1.3.)
- *Community* m. – Medicine in which the doctor serves a population, as a population. (Section 4.1.3.)
- *Evidence-based* m. – Medicine in which gnosis is informed by evidence from research (by its personal, critical examination by the doctor). (Section 4.1.5.)

- *Flexnerian* m. – Medicine in which gnosis is sought by means of the scientific way of thinking (learned from medical scientists working in laboratories). (Section 4.1.5.)
- *Knowledge-based* m. – Medicine in which gnosis is based on knowledge about gnostic probabilities. (Section 4.1.5.)
- *Rational* m. – Medicine in which the theoretical framework is rational. (Section 4.1.4.) See Scientific m. (below).
- *Scientific* m. – Medicine in which the theoretical framework is rational and the knowledge-base (of gnosis) derives from (quintessentially applied medical) science. (Section 4.1.5.)
- *Thinking-based* m. – Flexnerian m. (above).

Pathognomonic (as a quality of an element in, or entirety of, a diagnostic profile) – Conclusively indicative of the presence, or absence, of a particular underlying somatic anomaly (illness-definitional) or a particular extrinsic (or congenital) direct cause.

Patient (of a doctor in clinical medicine) – A person under a doctor's care on account of suffering from sickness (due to an existing case of its underlying, hidden somatic anomaly or an extrinsic direct cause; Section 3.3.1) or from an overt somatic anomaly (traumatic, say).

Prediction (in general) – Forecast. (Section 4.4.) See Forecast, Prognosis.

Prevention (in medicine) – Action intended to forestall future occurrence of an illness. See Prophylaxis.

- Primary p. – Prevention directed to first occurrence of the illness.
- Secondary p. – Prevention directed to recurrence of the illness.

Prognosis (in clinical medicine) – Gnosis about the future occurrence/non-occurrence of a particular event/state of health in a client (incl. about the way this depends on the choice of future treatment and/or lifestyle). (Section 4.4.) See Gnosis, Client, Patient, Forecast, Prediction, Treatment.

Prophylaxis (in clinical medicine): See Treatment.

Quackery – Pseudomedicine of the type in which possession of the relevant medical knowledge is merely an affectation. (Section 4.1.6.) See Medicine, Knowledge, Knowing.

Surgery (as an aggregate of disciplines of medicine, rather than as a modality of treatment) – The arts/disciplines/'specialties' of medicine in which the doctor, guided by his/her first-hand gnosis about the case, may execute a surgical treatment. (Section 4.1.2.) See Medicine, Treatment.

Treatment (in clinical medicine) – Action to change the course of a client's health for the better or to merely ameliorate their sickness. (Section 4.6.1.) See Intervention.

- *Prophylactic* t. – Treatment intended to prevent – to reduce (incl. eliminate) the risk of – the occurrence (prospective) of an illness or a sickness not due to illness. (Section 4.6.1.)

- *Palliative* t. – Treatment intended to ameliorate a sickness per se (without it being a therapeutic means to this end). (Section 4.6.1.)
- *Therapeutic* t. – Treatment intended to change the course of an illness for the better. (Section 4.6.1.)
- *Rational* t. – Treatment in the framework of a rational theoretical framework for prognosis with due regard for the state of the knowledge-base for the prognoses in the type of case and setting at issue and with due appreciation of the patient's valuations and preferences. (Section 4.6.2.)

Part III
General Principles of Medicine

Chapter 5
Logical Principles of Medicine

Contents

5.0 Abstract

While concepts are a prerequisite for any thinking and sound concepts are deployed in correct thinking, *principles* are express specifications of correct thinking – though possibly incorrect in this. (An 'orthodox' principle can be untenable, logically and/or ethically.)

General principles specific to (thought in) medicine can be derived on the basis of sound concepts together with select general dictates of two fundamental lines of scholarship (both aspects of philosophy): *logic* and *ethics*.

Much more than logical thinking and high ethics might well be expected of a doctor's practice of medicine, notably substantive mastery of the art (s)he practices in addition to those two virtues. But actually, a doctor's substantive competence can be seen to be a logic-dictated ethical imperative (Section 6.3).

© Springer International Publishing Switzerland 2015
O.S. Miettinen, *Medicine as a Scholarly Field: An Introduction*,
DOI 10.1007/978-3-319-19012-9_5

Logical principles of medicine have to do, principally, with thinking about the very core of medicine – *gnosis* in it (Section 4.1.1). These principles are general insofar as they are generic and, thus, without reference to any particular discipline of medicine.

5.1 Some Overarching Principles of Medicine

My dictionaries of medicine (refs. 1 & 2 in Section 3.1) define *principle* as:

> 1. a chemical component. 2. a substance on which certain of the properties of a drug depend [ref. 1];

and as:

> 1. A general or fundamental doctrine or tenet. See also law, rule, theorem. 2. The essential ingredient in a substance, especially one that gives its distinctive quality or effect [ref. 2].

These definitions are followed, respectively, by 9 and 27 particular principles. None of these exemplifies "A general or fundamental doctrine or tenet," specifically in medicine or otherwise.

While concepts are prerequisites for any thinking (Section 3.1) and sound concepts facilitate sound thinking – they were sought and proposed in Chapters 3 and 4 above – principles I take to be express *specifications of correct thinking.* And as with general concepts of medicine, I take *general* principles of medicine to be such specifications without reference to any particular discipline of medicine; that is, such cognitive guidelines relevant to all genuine, gnosis-centered disciplines of medicine. (Section 4.1.1.)

Some general principles were invoked in Chapters 3 and 4 already. The pre-eminent one not specific to the concerns in medicine was the logical guideline for the definition of a concept: it is to express the essence of the thing – entity, quality/quantity, or relation – at issue, meaning that which is true of all instances of that thing and unique to it; and to this end it is to specify the proximate genus of that thing together with the specific difference of the thing/species at issue within that genus. (Section 3.1.)

The preeminent, logic-based principle specific to medicine also came up. It was the tenet that, while the practice of medicine inherently entails the doctor's pursuit of esoteric knowing about the health of his/her client, attainment of this *gnosis requires bringing general medical knowledge to bear on the available facts on the case.* (Sections 4.2, 4.3, and 4.4.) And important principles of *causal thinking in medicine* were set forth in Section 3.6.

The principal ethical guideline in regard to gnosis also came up, having to do with the still-common situation in which the requisite general medical knowledge is nonexistent or otherwise unavailable: *the doctor is to acknowledge that (s)he doesn't possess the relevant knowledge, instead of affecting possession of the knowledge* (à la quackery). (Sections 4.1.6 and 4.5.1.)

5.2   The Requisite Knowledge for Diagnosis

The logic of diagnosis, as for the fundamental issue of *the nature of the requisite general knowledge* for it, is simplest for those instances in which the client is seeking certification of the absence of a particular illness, for the purpose of employment, for example. For such a *prefocused 'rule-out' diagnosis*, needed to know is, simply, what set(s) of facts would be *pathognomonic* about – conclusively indicative of – the absence of the illness in question. If the ascertainment of the facts on the case (i.e., on the diagnostic indicators involved) produces such a known (negatively) pathognomonic profile, 'rule-out' diagnosis about the illness is justified. But if not, the need is to know the non-zero level of the *probability* that the illness at issue actually is present (or absent).

If the diagnostic profile, in this situation or otherwise, is not pathognomonic about the presence (or absence) of the illness at issue, the question of principle is, How is one to generally think about the probability that the illness is present? In other words, the question is, What is that probability – the *correct probability* – in principle? To answer this question, one is to think about cases like the one at issue – instances of the diagnostic profile at hand – in general, in the abstract. In instances of this profile in general, the illness in question has a certain prevalence P, meaning that the illness is present in proportion P of these instances. The correct probability of the illness being present in the case at hand consequently has this same value, P. The need is to know about the level of this prevalence/probability.

The context for prefocused diagnosis can alternatively be one of *screening*, for a particular cancer, say; but in this context the concern ultimately is to *detect the latent presence* of the illness, insofar as it indeed is present. In a particular one of the repeat rounds of the screening, given a positive result of the initial test in it and upon the application of whatever further tests are prompted by this result, the need again is to know the probability of the presence of the illness at issue. For, this is needed for the choice among three options for action: (1) termination of the pursuit of the detection in this round of the screening (given sufficiently low probability); (2) added testing in this round of the screening (given an intermediate probability); and (3) proceeding to (early, latent-stage) treatment of the illness (on the premise of practical 'rule-in' diagnosis about it).

In both of these prefocused types of diagnostic pursuit the diagnostic indicators address the risk for the illness occurring (age, i.a.) and also potential manifestations – signs – of its presence in the results of laboratory testings (radiologic, pathologic, and/or other). The indicators do not address sickness from the illness (which, if present, remains latent), but for screening diagnosis the test results may need to include those from the most recent prior round of the screening (together with the timing of this).

. . .

In the much more common and generally more challenging type of situation for diagnostic pursuit, the doctor's client is a patient, presenting with a complaint about some *sickness*. Now the challenge is to achieve diagnoses about the possible proximal causes of this case of that sickness – ideally a 'rule-in' diagnosis about an underlying illness (somatic anomaly; Section 3.2.1) or identification of an extrinsic direct cause (e.g., the use of a medication; Section 3.3.1). For this, the point of departure generally is, specifically, the patient's *chief complaint* together with his/her *demographic category* (broadly). Quite analogous to sickness as a prompting of the diagnostic pursuit is an *abnormal test result* as an incidental finding.

Given the patient presentation in these terms and the premise (tentative perhaps) that the sickness (or the incidental laboratory finding) is not only genuine but due to an underlying illness (somatic anomaly), the doctor's first-order need again is to know *what set of items of information* is to be acquired for the diagnostic profile in the context of which the diagnostic probability-setting first becomes a timely concern, for the clinical diagnoses (Sections 4.2.3 and 4.5.2), that is. This is the counterpart here of the set of items of information that is needed for setting the diagnostic probability for the illness of prefocused concern above, for the first-stage diagnosis about it. But now the definition of the set of diagnostic indicators is more challenging, as the need is to discriminate among a set of possible underlying illnesses. The set of possible illnesses at the root of the chief complaint (or abnormal finding) – the differential-diagnostic set (Section 4.2.2) – determines the set of diagnostic indicators, the realizations of which are to constitute the patient's diagnostic profile for the first-stage, clinical diagnoses.

There is a tendency (among 'clinical epidemiologists,' notably) to think that the correct diagnostic probability in this general situation cannot have a general value, that the correct probability depends on the type and place of the practice. But in point of fact, a patient having sprained an ankle and having a radiograph of it should, ideally at least, be diagnosed with the same probability of tibia fracture by a doctor of whatever discipline in whatever type of practice and wherever. And a patient presenting with a complaint of recurrent fever would ideally be diagnosed with the same probability of malaria in Brussels and Brazzaville, with appropriate history-taking providing for this.

Once the information on the preordained set of diagnostic indicators has been acquired, the next need is to know what the complete differential-diagnostic set actually is, as the need is to achieve diagnosis about each of the illnesses included in this set; and the need intimately related to this is to know the profile-conditional probability of the presence of each of these illnesses (Section 4.2.2). If these probabilities are not sufficiently extreme (discriminating), the need is to know what added information could provide for this, and how the probabilities are to be set upon addition of these elements to the diagnostic profile.

. . .

The pivotal principle bearing on that common type of diagnostic challenge arguably is *the need to know the complete differential-diagnostic set*, notably when the case has turned out to be frustrating, with no rule-in diagnosis achieved and treatments regularly failing.

This is illustrated by an eminent book on doctors' thinking (ref. 1 below) – yes, *thinking* rather than deployment of knowledge (cf. Flexnerian medicine in Section 4.1.5). The book begins (p. 1) with the case of a patient who for 15 years had been sick, and ever more seriously so; had been seen by dozens of doctors of all conceivably relevant disciplines of medicine; and had been treated, quite unsuccessfully, for the various illnesses she had been diagnosed to have (irritable bowel syndrome, bulimia, anorexia nervosa, etc.; p. 14). Ultimately, yet another doctor thought of an illness that others had not considered (celiac disease; p. 15), proceeded readily to 'rule-in' diagnosis (endoscopic) about it, and the patient's recovery began with treatment directed to the patient's actual illness.

Reference 1: Groopman J. *How Doctors Think*. Boston: Houghton Mifflin Company, 2007.

Addressing this case in retrospect, Groopman explains (p. 24) that: "The doctors didn't stumble because of their ignorance of clinical facts; rather, they missed diagnoses because they fell into cognitive traps." And he explains more generally, though still as a matter of orientation only (p. 35), the nature of these traps. "Research shows," he says, "that most doctors quickly come up with two or three possible diagnoses from the outset of meeting a patient – a few talented ones can juggle four or five in their minds. All develop their hypotheses from a very incomplete body of information."

Groopman's own, first misdiagnosis had to do with "a middle-aged woman with seemingly endless complaints whose voice sounded to me like a nail scratching a blackboard. One day she had a new complaint, discomfort in her upper chest" (pp. 24–25). He never thought of dissecting aortic aneurysm as a possible cause of that sudden discomfort, and the outcome of the case was death from that illness. This case is, to him, another example of those "cognitive traps" – the trap of not thinking of all of the illnesses that are possible causes of the patient's sickness.

These examples illustrate Groopman's overarching idea that, in diagnosis, "most errors are mistakes in thinking" (p. 40).

Despite these authoritative ideas (of a Harvard professor of medicine) about mistakes in diagnostic thinking – in hypothesis-generation in particular, as part of the 'scientific method' in the pursuit of diagnosis (à la Flexner; Section 4.1.5) – I continue to hold the principle I stated above: *the diagnostician needs to know (sic) the complete differential-diagnostic set* that the patient's presentation (if presumed to be genuine and manifesting illness) implies. In both of those examples above, it was knowable that the illness whose presence was missed did belong in the complete differential-diagnostic set, and it therefore should have been targeted for diagnosis (among the others in the set). And in both of them the pursuit of diagnosis failed on the basis of not knowing the differential-diagnostic possibility that actually was the reality.

· · ·

I continue to hold, also, that the *diagnostician needs to know the profile-conditional probability* for each of the illnesses in the differential-diagnostic set. In this respect, a fundamental error/mistake is, I say, thinking that an alternative to this knowledge is what Groopman describes as: "Bayesian analysis, a mathematical approach to making decisions [*sic*] under uncertainty" (p. 61). In terms of this mistaken principle (cultivated by 'clinical epidemiologists'; Appendix 1), one "calculates a numerical probability for each possible diagnosis" (p. 62).

Groopman describes a case in which the diagnostician wanted to do this "Bayesian analysis," but there was "no database to refer to"; and so, he "tried to approach the problem not by sophisticated mathematics but by common sense" (p. 62). The latter the diagnostician took to be deference to the opinion of a consultant pediatrician, as the patient (in a hospital's emergency department) was a 10-year-old; and he took this deference to be a dictate of common sense even though he found the consultant's diagnosis to be quite unbelievable.

True common sense would have called for an inquiry into the basis of that surprising opinion, to distinguish between genuinely knowledge-based diagnosis and mere affectation of this in the case at issue (à la quackery; cf. Section 4.1.6). As it turned out, the consultant's diagnosis was very incorrect; and only acknowledgement of not possessing the requisite knowledge would have been consistent with freedom from the pretenses of quackery (Section 4.5.1).

. . .

When serious, expertly efforts at reaching diagnosis lead to no 'rule-in' diagnosis explanatory of the patient's sickness, there is a tendency to make the mistake of thinking that the efforts have failed to identify the illness that actually is at the root of the patient's sickness, and that it therefore is necessary to just accept the frustrating fact that the underlying illness remains unidentified. A particularly instructive example of this cognitive error is given in a book on a grand master of modern cardiology (ref. 2 below).

Reference 2: Lee TH. *Eugene Braunwald and the Rise of Modern Medicine.* Cambridge (MA): Harvard University Press, 2013.

A sister of Braunwald's mother was suffering from a serious case of lack of appetite and (its consequent) loss of weight. She had been seen by several leading physicians in New York City, including Braunwald's role model at Mount Sinai Hospital, but no cause for her worsening sickness could be found. Then, "One day, during their almost daily phone calls, Braunwald's mother pointed out [to him] that her sister was taking digitalis, the same cardiac medication that Braunwald had been studying for several years. 'Could this be responsible?' Claire Braunwald asked her son." (p. 175.) Thus coming to think of it, Braunwald was suddenly aware that his aunt's sickness indeed could be due to her use of digitalis. She was advised of this and asked to discontinue the medication's use, and "within two weeks her health had returned." (p. 175.) (Actually, it was her wellness that had returned, as she had recovered not from an illness but from a case of an exogenous sickness; Section 3.3.1.)

The morale of this story is an important principle: When diligent, expertly efforts at identification of the presence of an illness explanatory of the patient's sickness have failed, the need is to try to identify an extrinsic proximal cause of the sickness; that is, to shift from the diagnostic pursuit to its *etiognostic* alternative (in the effort to explain the sickness).

5.3   The Requisite Knowledge for Etiognosis

The example above has to do with a particular generic type of etiognostic challenge: needing to know about the etiogenesis of a case of *sickness* as a matter of its proximal/direct *extrinsic* causation (in the absence of an intrinsic one in an illness-definitional somatic anomaly). Another type of etiognostic challenge is the concern to know about the causal origin (proximal) of the *illness* that has been diagnosed as the cause of the sickness.

As that example illustrates, for the former type of etiognosis the critical need can be to know what antecedent(s)/concomitant(s) of the patient's sickness, known to have been present, could have been causal to the sickness; but alternatively, the need is to know what the known extrinsic causes of the sickness are in general and then to focus on the ones, if any, for which the patient has a positive history. And as that example also illustrates, the identification of a possible extrinsic cause of the case, there may be no need to consider the probability that it actually was causal: it may be possible, and commonly is, to simply remove the possible cause and to see whether recovery from the sickness results.

For etiognosis about a case of illness, insofar as this arises as a concern – for Workers' Compensation, say – there commonly is an a-priori specification of what a potentially etiogenetic antecedent of the case was; and for etiognosis about this in the case at issue, the need is to know about the probability of this antecedent having been causal to the case.

Analogous to this is the situation in which the illness (acute) may be 'iatrogenic' as a matter of having been caused by the use (recent) of a prescribed medication, and so also is the counterpart of this in respect to the use of an 'over-the-counter' type of medication (for which prescription is not needed). Etiognosis in these types of situation may be critically relevant for the prevention of the recurrence of the illness, apart from its potential relevance for dealing with the case at hand.

For etiognostic probability-setting in a case of illness, the case is to be specified not only in terms of the type of illness and the potentially causal antecedent of this together with the specified alternative to this in the causal contrast. To be included in the etiognostic profile of the case also are known modifiers of the level of the probability in question (i.e., modifiers of the causal rate-ratio of its occurrence), the realizations of these in the case at issue. Analogously with correct diagnostic probability, the *correct etiognostic probability* is the proportion of instances of the case profile in general such that the correct answer to the gnostic question is affirmative, here meaning that the antecedent actually was causal to the case (in the meaning specified in Section 3.6.2). This is, of course, the ultimate object of the knowledge required for etiognosis about a case of an illness.

5.4   The Requisite Knowledge for Prognosis

Like diagnosis and etiognosis, prognosis too can have a singular focus as to what the gnosis is about: prognosis can be about the possibility of coming down with a particular illness or sickness, say (an overt case of) a particular type of cancer or dementia.

For prognosis of this a-priori-focused type the need is to know, first, what set of prognostic indicators would be the basis of documenting a reasonable – suitably discriminating and practicable – prognostic/risk profile of the person (at the time of the prognostication). Apart from these factual inputs there may be the contingent ones of future lifestyle and/or prophylactic treatment. And then, conditionally on all of this, the need is to know the prognostic probabilities for the illness or sickness, specific to various points in or periods of prospective time – and possibly also for the entities involved in the possible side-effects of whatever prophylactic treatment is considered or actually adopted.

When the context for prognosis is the usual one – that of an existing case of a particular type of illness – there generally is no singular focus for what the prognosis is about, far from it. Commonly, various aspects of possible sickness from the illness need to be considered, and various possible complications of its course (Section 3.7.2) likewise. There may also be outcomes of the course (Section 3.7.3) of the illness to consider. And whatever treatments (Section 4.6.1) are entertained as possibilities, the various entities representing their potential side-effects may need to be considered.

The prognostic indicators in this context may well need to be known separately for each of the entities of health that the prognoses are about, and important among the prognostic indicators may be particulars of the illness and/or its manifestations at the time of the prognostication. All of this spells potentially great *complexity* for the requisite knowledge-base for the prognoses, a potential multitude of them in a given context.

A point of particular note in the context of an existing illness is this: The intervention-conditional prognoses are of interest as such, for adaptation to them; but their main utility has to do with assessment of the *relative merits of the available options for the treatment* and, thus, with the knowledge-based choice among them.

5.5   The Genesis of Gnostic Knowledge

A doctor's gnosis (dia-, etio-, or prognosis) about the health of a client is esoteric
on the ground that this knowing – particularistic – is based on bringing medical
knowledge – general, abstract – to bear on the available facts on the case. The
knowledge deployed for gnosis is, of course, specifically *gnostic knowledge* –
medical knowledge of the type that is fundamental to medicine, knowledge the
deployment of which is in the very essence of medicine (Section 4.1.1). In its
ultimate essence, as set forth in the Sections above (5.2, 5.3, and 5.4), this
knowledge is about probabilities conditional on the available sets of relevant facts,
on gnostic profiles of the cases encountered in (clinical) medicine. (Cf. Section 2.4.)

With the arts of medicine knowledge-dependent in this sense, understanding of
today's medicine involves possession of a suitable answer to the question, How has
the knowledge-base (gnostic) of medicine come about? that is, What has been the
genesis of the knowledge-base of medicine up to now? And the associated normative
question naturally is: How should the knowledge-base of medicine be advanced in
the future, to make future medicine more advanced than the medicine of the present?

For meaningful answers to these questions it is essential to appreciate, first, the
complexity – the highly multifarious contents – of the requisite knowledge-base
of modern medicine (Section 1.5). Even in quite narrowly-defined disciplines of
medicine there commonly is a substantial number of types of case presentation for
a given species of gnosis, and in the context of a given one of these presentations
there generally is an enormous number of possible gnostic profiles conditionally on
which the doctor may need to know about a number of probabilities (gnostic).

And it is necessary to appreciate, also, that a doctor's learning about any
given gnostic probability on the basis of experience in practice – this '*case-based
learning*' about these (supplementing their medical-school experience with this;
Section 2.4) – is wrought with serious problems. The individual cases, if instructive
at all, are not ones of established probabilities but only instances of getting to
know the truths whose probabilities are the gnostic concerns. Thus, a doctor's case-
based learning about any given profile-conditional probability can only be based
on experiences of this type with a variety of other profiles (as any given profile
generally is extremely uncommon in a doctor's experience).

Such learning obviously is an *enormous cognitive challenge*, even in the context
of a perfect recollection of all of the instances that in principle have been informative
of the probability at issue. (More on this in Section 8.1.)

For today's medicine the knowledge-base (for setting profile-conditional gnostic
probabilities), such as it is, has in all essence been derived from personal experiences
in the practices of doctors, rather than from gnostic research (Section 1.4). And in
consequence of this, gnostic probabilities – the enormous multitude of these that
are the concern in any given doctor's practice of knowledge-based medicine –
are nowhere expressly codified but are, merely, matters of '*tacit knowledge*' of
doctors. This 'knowledge' takes manifestation only on a case-by-case basis; and the
case-specific probabilities commonly are quite discordant across individual doctors,

even among 'experts' on commonly-needed gnostic probabilities in emergency and intensive-care medicine.

This state of affairs is not inherent in medicine, but progress toward genuinely knowledge-based medicine will require extensive *gnostic research* in a logically tenable theoretical framework of medicine itself and, consequently, of the formulation of the objects – and secondarily the methods – of this research.

I give a brief introduction to these theoretical matters in respect to diagnosis in Appendices 1 and 2. A much more comprehensive treatment of gnostic research can be found in my book *Toward Scientific Medicine* (Springer, 2014).

5.6   The Accessibility of Gnostic Knowledge

For medicine to actually be knowledge-based (which is said to distinguish it from quackery; Sections 4.1.4 and 4.1.6), the first requirement naturally is that the requisite knowledge (for gnostic probability-setting) must exist. Second, the knowledge must be accessible to the doctor when it is needed, quite urgently perhaps. And third, the doctor must defer to the knowledge and not override it by the deployment of personal opinion in lieu of the knowledge.

In the foreseeable future, insofar as there gets to be commitment to the cultivation of genuinely scientific medicine (Section 4.1.5), the existence of the requisite knowledge – uncertain and dated perhaps, but genuine – will presumably become quite commonplace. And before the advent of such scientific knowledge, experts' tacit 'knowledge' may become codified in terms of the typical levels of their profile-specific gnostic probabilities, and these may serve as ersatz knowledge for doctors at large (Section 4.5.1).

In that future, accessibility of the existing gnostic knowledge will be assured by the availability of '*expert systems*' that bring the needed knowledge to doctors, and right where and when it is needed. (Appendix 3.)

5.7   The Deployment of Gnostic Knowledge

Insofar as directly practice-relevant knowledge (gnostic) has been achieved and ad-hoc accessed, its first-order deployment naturally is its application to the gnostic profile of the case at hand, to derive (at least some of) the gnostic probabilities that generally are of concern in the context of the profile.

The second-order deployment of the knowledge is, just as naturally, application of the gnoses it provided for. On this level of deployment the overarching principle is quite obvious: since the gnoses represent the doctor's esoteric knowing about the health of the patient or other client, they are deployed in 'doctoring' the patient/client in the etymology-based meaning of 'doctor': teaching the client about their own health (insofar as this is possible in the context). (Section 2.1.)

Once the client thus gets to reach essentially the same level of knowing about their own health as the doctor has about this in terms of the gnoses, the next-order deployment of the gnostic knowledge generally has to do with *decision-making* in the light of the gnoses, concerning possible further testings or choice of treatment. The principle that is fundamental to medical decision-making is this: The optimal *choice among the options for action is not implicit in the gnoses bearing on the choice; its optimum depends, also, on the valuations – the perceived utilities/disutilities –* of the medical and other implications of the possible actions, and th*ese are not matters of medical knowledge* but are, instead, generally particular to the client/patient. From this flows the *core principle of medicine*: Decision-making about the choice of action is a prerogative of the doctor's gnostically informed client, not the doctor's own. (Sections 6.2 and 6.3.)

All of this implies a fundamental principle concerning *guidelines for practice*: to whatever extent normative, they must address gnostic knowledge and not choice of action informed by the knowledge. Current practices in the promulgation of practice guidelines are, however, much at variance with this principle. For one, they are based not on knowledge but merely on available evidence; and that evidence generally does not involve the necessary distinctions among the various possible gnostic profiles. Instead of representing guidelines as to what the knowledge might be taken to be, they generally are guidelines/recommendations – by the reviewers of the evidence themselves – for the choice of action! (Cf. rational medicine in Section 4.1.4.)

5.8 Economists' Principles for Medicine

5.8.1 The Economic Aspect of Medicine

Medicine has an aspect such that, while it isn't in the essence of medicine, its importance has come to have a tendency to trump all that is intrinsic to the professions of medicine: the *economic* aspect of healthcare by doctors together with their allied, paramedical professionals. Arguably at least, medicine as a societal institution, especially subsequent to the advent of national health insurance (Medicare for the elderly and Medicaid for the indigent in the U.S.; universal Medicare in Canada), has become more a problem in itself than a solution to problems extrinsic to it (those addressed in Chapter 3), as its cost to society is felt to be excessive in relation to the benefit from it.

Anything produced in the 'productive arts' of medicine (Section 4.1.1), whatever may be its medical utility/disutility (to the client and others), is generally produced on various levels of efficiency and, thereby, of economy; and the *economic aspect of a given instance of task execution in medicine thus is its degree of efficiency,* conditionally on the utility/disutility of its product. Where an element of medicine is useless or counterproductive, it thereby is of objectionable quality in its medical aspect; but even such elements of medicine, where they occur, should be of good quality economically: the result – none or harm/disutility – should be achieved economically, just as any benefit/utility should be. Ideally, of course, the quality of medicine is good in both the medical and the economic dimension of it.

With medicine such a vast industry in a modern society and, hence, such a large segment of any modern society's economy, it has attracted the attention of *economists* despite the esoteric nature of these services, insurance coverage of the costs of these, and other peculiarities in the framework of service industry at large. This attention has been, broadly, of two very different kinds, corresponding to a fundamental duality in economists' conception of what their field is about.

5.8.2   The Duality in Economists' Outlook

The Catallactic Outlook Traditionally, economics has been a field concerned with the *workings of free markets*, as to the implications of supply and demand on pricing, etc. 'Catallactics' is a synonym of 'economics' in this traditional meaning of 'economics.' A *fundamental tenet* in catallactics is that free markets enhance the substantive quality of goods and services, in addition to determining their costs – reducing the latter by competition.

In the catallactic spirit, two economists recently published an ambitious book (ref. below) about a needed major innovation in healthcare, aimed at enhancing both the medical and the economic quality of the care provided. The authors see the need to *arrange for productive competition* among the various providers of healthcare, specifically in terms of *"outcomes" of the care, mandated to be routinely measured and reported.*

Reference 1: Porter ME, Teisberg EO. *Redefining Health Care. Creating Value-Based Competition on Results.* Boston: Harvard Business School Publishing, 2006.

. . .

The Praxeologic Outlook In a more modern conception of it, economics is praxeology – the science or theory of human action in general, as most eminently elaborated by Ludwig von Mises (ref. below).

Reference 2: Mises L v. *Human Action. A Treatrise on Economics.* Third revised edition. Chicago: Henry Regnery Company, 1963.

In this framework, actions in healthcare are a concern of economic theory of purposive behaviors, actions flowing from decisions about them. This has lead 'health economists' to concern themselves with *medical decision-making* in the framework of *decision theory*. This theory was originally a purely statistical one (of Neumann & Morgenstern) of 'decision-making under uncertainty,' with the aim of maximizing 'expected utility.' It was adopted for decision-making in business and then adapted to medical decision-making by 'health economists.'

A centrally important, *fundamental precept* of von Mises was this:

[Praxeology] is a science of the means to be applied for the attainment of ends chosen. . . . Science never tells a man how he should act; it merely shows how a man must act if he wants to attain definite ends. [P. 10]

5.8.3 Economists' Principles for Medicine

Health economists' teachings about decision-making in medicine are quite different from my precepts in this book. I have thus far focused on the expressly medical aspect of doctors' healthcare, and my decisions-oriented main points in this have been that a doctor is to teach the client about their health (Section 2.2.1) – naturally inclusive about how it could be changed for the better – and that the appropriate decision-making about actions is not a prerogative of the doctor but of the client (Section 5.7). As the focus now is on the economic aspect of the care, I revise the former point by saying that the doctor needs to teach the client not merely about their health (inclusive of the health implications of possible treatments) but also about the costs of the care actions being considered – costs to the insurer (society and/or other) as well as to the client personally.

Health economists tend to focus on decisions in the adoption of some medical *intervention*; and in this, they generally take it as a given that in any medical intervention the "end chosen" for it (cf. von Mises above) generally is *increase in the client's quality-adjusted duration of life*, where quality is *health-related quality of life* (HRQL).

That HRQL is assigned value 1 in reference to 'perfect health' in the meaning of absence of the health problem at issue; and the minimum is set to 0, corresponding to 'death' (with the meaning of 'death' as a level of HRQL – of life! – left unspecified). The level of a particular person's HRQL is assessed by means of economists' 'standard gamble,' which addresses the extent to which risk of immediate death from a (notional) treatment would be acceptable on the premise that the treatment otherwise is fully curative of the illness (or sickness) at issue.

This thinking about medical decision-making has spawned much 'health services research' to quantify the HRQL implications of various chronic illnesses (as though those implications were universal).

This presents an insurmountable intellectual challenge to me – most broadly the overarching idea that, under economists' principles, a doctor ideally 'calculates' whether the client should be treated in a particular way, in the client's personal interest or 'from the societal perspective.' For there surely are states of health such that the overall HRQL is negative, so that status post mortem (the state of being dead) is preferable to the person in question (and assisted suicide thus is desirable to them); and the nonnegative HRQL concept of 'health economists' thus seems not to be philosophically 'real' and, thus, admissible. The risk tolerance for immediate death in the essence of the 'standard gamble' has major determinants other than HRQL – economic and ethical ones, for example.

· · ·

These technicalities aside, a salient problem with these teachings is their being in violation of a core principle of praxeology per se: Economics in this modern, larger meaning of the term has to do with actions in which each actor – *homo agens* – takes whatever decision about action to the end of their own happiness (ref. 2 in Section 5.8.2 above), consistent with what philosophers teach (Section 2.3). The decision-making homo agens in medicine properly is the doctor's client (Section 5.7, i.a.), and optimization of the care recipient's decision is not subject to any decision-theoretical calculation; it is incalculable.

This decision-optimization calculus came to medical academia via academic schools of business (and to these from statistics; cf. Section 5.8.2 above); and the decision theory at the root of it serves certain types of business decisions very well: Business decisions about service of a given type – to sell a certain type of insurance in a given type of situation, say – are recurrent, and the 'expected utility' to the actor (the company) actually materializes to the decision-maker (the company again) in the long run but it does not materialize to the clients of the business. Decision theory justifies a company's selling of life insurance, say; but purchase of this, while it may be perfectly rational, cannot be justified on the basis of its 'expected utility,' which is negative (in monetary terms).

When a doctor's client decides to accept whatever medical ministration, the decision generally is akin to buying life insurance: it is of the once-in-lifetime type, with no expectation to derive a statistical 'expected utility' from it, much less a positive one at that.

Health economists actually well understand this, and so they make the calculations 'from the societal perspective.' Yes, any given decision in clinical medicine is recurrent, practically ad infinitum, from the societal perspective; but now the problem is that *society is not the decision-maker* in clinical medicine (different from community medicine).

5.9   Epilogue

In this Chapter on principles "of" medicine, a distinction has not been clearly made between thinking *about* medicine – as when studying medicine – and thinking *in* medicine, when practicing it. Regarding thinking about medicine, the critically important, fundamentally orientational proposition here has been that in the essence of – that is, essential in – medicine is *knowing* (Sections 4.1.1 and 4.1.6) rather than thinking (Sections 4.1.5 and 5.2).

Consonant with and illustrative of this, many decades ago, a pioneering cardiovascular surgeon (John Kirklin) remarked to me that he considers it his ethical imperative to formulate and adopt his principles of surgical decision-making at his "brightest moments" (by thinking about them, critically) and then, in actual patient-care, to simply apply (unthinkingly) such a decision algorithm, independently of such potentially misleading influences (on thinking) as the outcome of his most recent operation of that type, his mood at the time of the decision, etc. And in line with this, his post-operative, recovery-phase care was maximally computer-driven and thus knowledge-based rather than thinking-based.

All this to underscore that practice of medicine indeed is, in its ideal form, not a matter of thinking – not thinking about the client nor about the options for what to do for the client. Instead, medicine ideally is *reflexive doing the right things* – from the initial clinical observations and history-taking all the way to the relevant gnoses as the basis of 'doctoring' (teaching) the client about their health. The more professional a doctor is, the less (s)he is a thinker and, thus, the more (s)he is a professional automaton.

This *fundamental truth about the logical principles of medicine* is difficult for doctors steeped in the prevailing culture of medicine to accept. But my textbook of tennis (Vic Braden, 1977) says that a champion in this game is not a creative player but, instead, someone who is proud of having learned to "keep on producing the same old boring winner." In medicine, due to its esoteric nature (Section 4.1.1), the clients generally are unable to identify who the masters of their particular disciplines are. Hence, successful practice of pseudomedicine (Section 4.1.6) is possible, while the masters in tennis are obvious and, therefore, successful practice of pseudotennis – unappreciative of the essence of the game and of the logical principles of it – is impossible.

It follows that keen concern to grasp not only the essence of genuine medicine but also the logical principles of practice that flow from this – and ultimately to practice accordingly – requires commitment to *ethical* principles of medicine (cf. master surgeon above). It is for reasons not of recognition and careerism but of ethics that a learned doctor provides the best possible care to his/her clients.

Chapter 6
Ethical Principles of Medicine

Contents

6.0 Abstract

A centrally important aspect of the quality of a doctor's practice of medicine obviously is its ethics. Preparatory to grappling with the principles of medical ethics is introduction to ideas about ethics in general and also to those about professional ethics more specifically.

The core principle of professional ethics specific to medicine can, I suggest, be deduced from the general essence of genuinely medical professions (Section 4.1.1) together with the Aristotelian principle of ethics – the pursuit of happiness by means of the virtue of cultivating *excellence* in oneself – applied to oneself as a doctor.

Against this backdrop it is of interest to examine the oath of physicians the formulation of which is (falsely) attributed to 'The Father of Medicine' (Section 1.1). Also, examination of major examples of *amoral* practices of medicine is instructive.

© Springer International Publishing Switzerland 2015
O.S. Miettinen, *Medicine as a Scholarly Field: An Introduction*,
DOI 10.1007/978-3-319-19012-9_6

6.1   Ideas About Ethics in General

'Ethics' and 'morals' are synonyms, etymologically Greek and Latin, respectively; *ethos* and *mores* both mean habits or customs in a community – normative conduct in this sense. Kant (ref. 1 below; p. 1) taught that, in philosophy, ethics has to do with "the conduct of beings possessed with free will" and that ethics is one of the two concerns of philosophy (the other being knowledge).

Reference 1: Kant I. *Lectures on Ethics* (Translated by Louis Infield, Foreword by Lewis White Beck). Indianapolis: Hackett Publishing Company, 1963.

In moral philosophy, the central question is, What is to be meant by human conduct being, or not being, moral? Before outlining philosophers' answers to this question, I note that in my Oxford American Writer's Thesaurus a *moral* man is characterized as "virtuous, good, righteous ... ," with 'dishonorable' given as the antonym. For 'immoral' this source gives as synonyms "unethical, bad, morally wrong, ..." And it points out that "*Immoral* means 'failing to adhere to moral standards' [and is] always judgemental," while "An *amoral* person has no understanding of these norms, or no sense of right and wrong."

That question about the essence of moral conduct has not been resolved by philosophers. Very broadly, it can be said that some of them have represented the *deontological* outlook, with concern to understand the essence of moral obligations immanent in 'moral laws,' while others have held the *teleological* stance to be fundamental to moral conduct, meaning that to them it is the actor's purpose or the actual consequences of the action that may or may not make the action morally justifiable.

The ethics of Maimonides (Section 2.1) was deontological, founded as it was on scriptural commandments (Mosaic Law) in the Torah. In Kantian ethics, correspondingly, "The dictates of moral imperatives are absolute and regardless of the end" (ref. 1 above, p. 5), with God the lawgiver (p. 56) though not expressly revealing the laws. Kant's formulation of the "supreme principle of morality" was this: "All action ... must be such that it can at all times become a universal rule; if it is so constituted, it is moral" (p. 43).

To Maimonides, the moral laws, even though deontological, actually had a teleological role – as the means to the end of becoming a knowing creature and attaining human perfection (Section 2.1). Indeed, he was quite in tune with Aristotelian ethics which, even though it involved no moral imperatives, was a philosophy of virtuous living in the meaning of cultivating one's personal characteristics to the end of personal happiness, the cultivation possibly leading all the way to magnificence (of which he himself, arguably at least, was a paragon). And Kant, correspondingly, held that "The ultimate destiny of the human race is the highest moral perfection, provided that it is achieved through human freedom, whereby alone man is capable of the greatest happiness" (p. 252). Both Maimonides and Kant stressed the importance of the right motive in moral conduct – "an inner impulse to give expression to the moral law on account of the inner goodness of such action" (p. 41).

Purely teleological is the moral philosophy called *utilitarianism*. In it, the core principle is that an action is morally right or wrong according as it tends to promote, or to take away from, the aggregate happiness of those affected by it. While deontological morality involves a wholesome will in the execution of an obligation, in utilitarianism it is the action alone that has, or does not have, moral justifiability. Different from deontological morality, value judgements are involved, centrally, in utilitarian ethics. This, however, entails considerable theoretical problems in this line of moral philosophy.

Utilitarian conduct is commonly viewed as altruistic, in contrast to egoistic conduct. But, arguably at least, all of human action, utilitarian action included, is directed to enhancement of the actor's personal happiness (cf. Maimonides and Kant above, and Sections 2.3 and 5.8.2).

6.2   Ideas About Professional Ethics

Before focusing on ethics specific to the medical professions, it is relevant to take note of ideas about ethics in professions in general. As a guide to this I use the only text on this topic I have on my shelves (ref. below). Under its dedication statement up front there is a notable quote from Cicero: "To everyone who proposes to have a good career, moral philosophy is indispensable."

Reference: Callahan JC (Editor). *Ethical Issues in Professional Life*. New York: Oxford University Press, 1988.

In the Preface the Editor points out that "a number of college and university philosophy departments have recently introduced nonspecialized courses on professional ethics," adding that "This book is designed to serve as a text for such courses." And she points out that "questions pertaining to the moral [*sic*] foundations of professional ethics are intimately connected to questions about the appropriate model of the *professional-client relationship* that are, in turn, intimately related to questions regarding client deception and informed consent, privacy and confidentiality, and the obligations of professionals to third parties and society at large" (italics mine).

From the vantage of the present introduction to medicine, it is notable that the Editor makes no point about the philosophy departments in colleges and universities having, or not having, introduced courses on ethics specific to particular professions, most notably the medical professions – taught to students preparing for careers in the professions. The implication may be that the essence of professional ethics is generic – the same across all learned professions.

Of the 54 articles in that book, only one (#14) addresses those pivotal "questions about the appropriate model of the professional-client relationship" in general, without specificity to any particular profession. It is an article by M. D. Bayles, entitled "The Professional-Client Relationship." In it, his purpose is "to develop an ethical model that should govern the professional-client relationship." Bayles says, for orientation, that *"The central issue in the professional-client relationship is the allocation of responsibility and authority in decision making* – who makes what decisions [italics mine]. The ethical models are in effect models of different distributions of authority and responsibility in decision making."

In this article, the subheadings are Agency, Contract, Friendship, Paternalism, and Fiduciary. And insofar as Bayles actually does "develop an ethical model that should govern the professional-client relationship," this model amounts to the following passage in the section on Fiduciary: *"The appropriate ethical conception of the professional-client relationship is one that allows clients as much freedom to determine how their life is affected as is reasonably warranted on the basis of their ability to make decisions."* (Italics mine.)

6.3   The Core of Medical Ethics

That article on the professional-client relationship does not define the fiduciary aspect of it. For that word, the etymology has to do with trust, and in any professional-client relationship there is a central role for trust. In learned professions, more specifically, *the professional has the ethical obligation to be worthy of the trust that the client places in him/her.*

It is in the essence of the doctor-client relationship that the client is 'employing' the doctor to serve as a representative of the medical professions. The client therefore is entitled to expect the doctor's service to represent *true professionalism* in the meaning of its being, in every respect, as good as can reasonably be expected from anyone in the profession relevant to the case in the setting of the care; the client should be able to trust that this indeed is the case. This means the *ethical obligation to provide excellent service*, to embody the Aristotelian virtue of excellence (Section 6.1) in the service. This obligation I take to be the *central ethical imperative* governing any doctor's professional work, the medical counterpart of Mosaic Law in Judaism (Section 2.1).

Just as Maimonides held that rabbinical laws need to be deduced – under express principles – from the canonical, Mosaic Law (Section 2.1), I suggest that the various particular ethical principles of medicine are to be deduced from that overarching ethical obligation in medicine. Rather than attempting to sketch many of the derivative principles here, I merely underscore what I take to be the two most proximal ones of these: (1) A doctor has the ethical obligation to be *committed to serve the client(s) alone*, with no compromise in this from other interests (furtive self-interests in particular); and (2) a doctor has the ethical obligation to be *medically competent* in his/her professional work, which means the obligation to limit the services (s)he provides to ones in which (s)he is as competent as anyone.

6.4   The Hippocratic Oath

The 'Hippocratic' oath, which actually was not written nor otherwise adduced by 'The Father of Medicine' (Section 1.1), still retains considerable status in ethics specific to medicine. This Oath is now commonly presented to be the following (incl. in refs. 1 & 2 in Section 3.1):

> I swear by Apollo the physician, by Aesculapius, Hygeia, and Panacea, and I take to witness all the gods, all the goddesses, to keep according to my ability and my judgement the following Oath:
>
> To consider dear to me as my parents him who taught me this art; to live in common with him and if necessary to share my goods with him; to look upon his children as my own brothers; to teach them this art if they so desire without fee or written promise; to impart to my sons and the sons of the master who taught me and the disciples who have enrolled themselves and have agreed to the rules of the profession, but to these alone, the precepts and the instruction. I will prescribe regimen for the good of my patients according to my ability and my judgement and never do harm to anyone. To please no one will I prescribe a deadly drug, nor give advise which may cause his death. Nor will I give a woman a pessary to produce abortion. But I will preserve the purity of my life and my heart. I will not cut for stone, even for patients in whom the disease is manifest; I will leave this operation to be performed by practitioners (specialists in this art). In every house where I come I will enter only for the good of my patients, keeping myself far from all intentional ill-doing and all seduction, and especially from the pleasures of love with women or men, be they free or slaves. All that may come to my knowledge in the exercise of my profession or in daily commerce with men, which ought not to be spread abroad, I will keep secret and never reveal. If I keep this oath faithfully, may I enjoy my life and practise my art, respected by all men and in all times; but if I swerve from it or violate it, may the reverse be my lot.

As for the contemporary implications of this Oath, a modern student of medicine, I suggest, might weigh and consider whether (s)he too could, and even should, be taught his/her discipline-to-be by a *single master* of this art. (S)he should be clear on what existing *rules of the profession*, if any, (s)he should, and should not, agree to. (S)he should appreciate that his/her ethical imperative in the practice of medicine is not merely to function according to his/her attained *ability and judgement* but to an objective, higher standard. And as for *the precepts and the instruction* in the education which (s)he has entered, (s)he should not, I propose, be that deferential to the teacher(s) – recall Section 2.4 in particular – but should plan to weigh and consider each of these as to their merits as contributions toward the ethical goal of excellence as a professional (Section 6.3 above).

I discuss these matters in Chapters 7 and 8 and in Appendix 3 in addition to Section 6.3 above.

6.5   Amoral Practices in Medicine

A doctor's professional actions are *amoral* – rather than unethical/immoral – if (s)he has no sense of right versus wrong, or if (s)he actually does have a sense of this and also acts accordingly but *lacks understanding of the nature of the ethical imperatives* – 'moral laws' – to which the practices of medicine should conform (Section 6.1).

Where a doctor's practices are amoral, they commonly are ethics-free in that latter, lack-of-understanding sense of this. And this ethical flaw is, specifically, consequent to not understanding the *objective* nature of the ethically-required excellence in medicine (Section 6.3). Instead of this understanding, the thinking commonly is that excellence in medicine is tantamount to the doctor's acting to the best of his/her ability and his/her own judgement (Section 6.4 above), such as they are. (Cf. philosophy underpinning Evidence-Based Medicine, in Sect. 4.1.5.)

When the requisite knowledge (Chapter 5) remains nonexistent, as is largely the case even in modern medicine, excellent 'doctoring' (teaching the client; Section 2.1) nevertheless is characterized by truthfulness of the doctorings. This entails, inter alia, clear acknowledgement of the nonexistence of knowledge about the relevant truth when this is the case, while affecting possession of that knowledge would be quackery (Section 4.1.6).

In the situations of called-for acknowledgement of agnosticism, a doctor runs the risk of *the hubris of substituting an unfounded subjective opinion for the nonexistent knowledge*. This hubris leads to amoral practice in a third meaning of this. Two examples of such practice follow, both tragic in the extreme.

. . .

Surgical Practices In his "biography of cancer" (ref. 1 below) Siddhartha Mukherjee gives, inter alia, an account of the evolution of treatments for this "emperor of all maladies." In surgery for breast cancer, the pioneer and authority got to be *William Stuart Halsted*. At Johns Hopkins as of 1889, he "attacked breast cancer with relentless energy" (p. 65). This surgery of his became ever more "radical," and hence, "the mammoth mastectomies permanently disfigured the bodies of his patients" (p. 65). Following Halsted's lead, "By 1898, [surgery] had transformed into a profession booming with self-confidence, a discipline so swooningly self-impressed with its technical abilities that great surgeons unabashedly imagined themselves as showmen. The operating room was called an operating theater, . . ." (p. 66).

Reference 1: Mukherjee S. *The Emperor of All Maladies. A Biography of Cancer.* New York: Scribner, 2010.

As survival rates of breast-cancer patients remained disappointing, "Halsted and his students . . . clutched to their theories even more adamantly," extending further the "cleaning out" of lymph nodes (p. 68). "'Radicalism' became a psychological trap, . . ., central not only to how surgeons saw cancer, but also in how they imagined themselves" (pp. 69–70). "This trajectory toward more and more brazenly aggressive operations . . . mirrored the overall surgical thinking of the early 1930s. . . . Surgeons often counted themselves lucky if their patients merely survived these operations" (p. 70).

"Undoubtedly, if operated upon properly the condition may be cured locally, and that is the only point for which the surgeon must hold himself responsible,' one of Halsted's disciples announced at a conference in Baltimore in 1931. ... Curing cancer was someone else's problem" (p. 70).

In St. Bartholomew's Hospital in London in the 1920s, Geoffrey Keynes had introduced "a careful mixture of surgery and radiation, both at relatively minor doses. ... But Halsted's followers in America laughed away Keynes's efforts" (pp. 195–196).

In treatment for cancer, "Radical surgery, Halsted's cherished legacy, had undergone an astonishing boom in the 1950s and 1960s. In surgical conferences around the world, Halsted's descendants ... had stood up to announce that they had outdone the master himself in their radicalism. ... The radical mastectomy had thus edged into the 'superradical' and into the 'ultraradical,' an extraordinarily disfiguring procedure in which surgeons removed the breast, the pectoral muscles, the axillary nodes, the chest wall, and occasionally the ribs, parts of the sternum, the clavicle, and the lymph nodes inside the chest" (pp. 193–194).

Leading surgeons, "all encaptured intellectual descendents of Halsted, were the *least* likely to sponsor a trial that might dispute the theory they had so passionately advocated for decades" (p. 198). But ultimately the needed study was carried out, and published in 1981. "The group treated with the radical mastectomy had paid heavily in morbidity, but accrued no benefits in survival, recurrence, or morbidity" (p. 201).

. . .

Psychiatric Practices Even though in the practice of surgery, unsurprisingly perhaps, there have been notable examples of the absence of the requisite intervention-prognostic knowledge leading to therapeutic hubris (as illustrated above), a beginning student of medicine scarcely would expect such hubris to have characterized the medical discipline concerned with *'mental illnesses'* (i.e., illnesses – somatic – manifest in mental sicknesses; Sections 3.2.2 and 3.4). But (s)he needs a reality check on this (ref. 2 below).

Reference 2: Scull A. *Madhouse. A Tragic Tale of Megalomania and Modern Medicine*. New Haven: Yale University Press, 2005.

A century ago, doctors concerned with mental sicknesses "were speculating in this species of human misery" and laying claim to "special powers to identify, to manage, or even to resurrect those victimized by the corruption and death of the mind" (ref. 2 above, p. 14). But it was with "veiled contempt [that] the rest of the medical profession viewed their psychiatric colleagues" (p. 21).

In this situation, some "bright and brash young men" on the staff of a mental institution – Worchester State Hospital in Massachusetts – set out to *"bring the tools and techniques of the new scientific, laboratory-based medicine to bear on the recalcitrant problem of madness,* dispelling etiologic ignorance and overcoming therapeutic impotence" (p. 21, italics mine). In this way, "the source of the madness would soon be lifted, just as surely as Koch, Pasteur, Lister, and their allies had triumphed over other forms of deadly and hitherto incurable disease" (p. 25).

"Though the medical profession had initially greeted the claims of Lister and Pasteur with skepticism and ridicule," gradually "the medical outlook seemed completely transformed. ... Potentially, the practical payoffs of the bacteriological revolution seemed limitless" (p. 28). And as for psychiatry, a promising paradigm was the syndrome of 'paresis' that got to be understood to be a manifestation of terminal-stage syphilis, subject to diagnosis and also treatment directed to the bacillus (pp. 30–31). This was extrapolated to the idea that mental sickness in general could result from a "hidden infection" somewhere in the body, just as leaders of medicine at large were suggesting for arthritis and various other chronic illnesses (p. 31).

This perception of the possibility that bacterial infection is the somatic anomaly at the root of whatever type of debilitating mental sickness soon transformed in the minds of some leading psychiatrists to the (apparent) *belief* that those sicknesses indeed are manifestations of hidden infection; and this belief became the basis of *treatment* in state-run mental institutions – surgical removal of body parts potentially harboring the madness-causing infection.

Another belief bearing on the treatment was that the patients' "madness leaves them bereft of their status as moral agents," which was taken to be a justification "for seeing them as *less than fully human*, and for treating them accordingly" (p. 13, italics mine). In an avant-garde institution for the practice of the treatments – Trenton State Hospital in New Jersey – this was taken to mean that "Protests from patients and their families ... must be ruthlessly pushed aside as short-sighted preferences reflecting the patients' mental disturbance and incompetence or the imperfect knowledge of their families" (p. 55) – implying that the psychiatrists had the requisite competence and perfect knowledge. The leader of this hubris – a counterpart of Halsted's 'radicalism' – was *Henry Cotton*, a thoroughly educated psychiatrist, at that hospital in Trenton.

These practices of that avant-garde hospital were in tune with the ideas in *the avant-garde of medical academia* regarding various chronic diseases. Thus, for instance, the dean of an eminent school of medicine (resulting from Rush Medical College having merged with University of Chicago) – Frank Billings, "one of the most prominent figures in early twentieth-century American medicine" (p. 32) – in his lectures in another eminent school of medicine (the Lane lectures at Stanford Medical School in 1915) "seized the occasion to advocate the vital importance of converting a new generation of physicians to the war on sepsis. In almost apocalyptic language, he urged his listeners to 'make sure that all sources of focal infection have been obliterated'" (p. 32).

"Such beliefs were not only the peculiar province of the medical elite of the Windy City [Chicago]. Similar views were held in the *bastions of early twentieth-century scientific surgery and medicine*: at the Mayo Clinic, ..., and at Johns Hopkins, ..." (p. 33, italics mine). Statistics on surgeries in some New York and Pennsylvania hospitals from the early 1920s "starkly revealed ... the penetration of the ideas associated with the doctrine of focal infection into the center of routine American medical practice" (p. 33). "Across the Atlantic, leading figures in British medicine and surgery hastened to add their voices to the chorus, ..." (p. 34).

In the avant-garde mental institution of the practice founded on this doctrine (specified above), "patients were all devoid of teeth, and they were not given dentures" (p. 135). In a 3-month period in 1925 in this institution, 54 patients underwent surgery, many of them with several types of surgery. In this period in this hospital on these 54 patients, "There were 45 appendectomies and 45 underwent Lane's operation, designed to release adhesions in the colon; 27 operations on the small intestine and stomach; 9 removals of ovaries and fallopian tubes; 3 hysterectomies; 4 colostomies; and a variety of other operations, from the removal of the spleen to the removal of hemorrhoids" (p. 201). Of these patients in the ensuing 5–8 months, 16 % died and another 40 % remained hospitalized – "mangled and disfigured without any offsetting therapeutic advantages" (p. 202).

While these were fragmentary observations of an outside psychiatrist sent to conduct a thorough investigation of the records in that hospital, *the full report's publication was blocked by the establishment*; it "remained safely buried in the archives at Hopkins and Trenton" (p. 290).

For the therapeutic hubris based on the focal infection doctrine, there got to be other, equally tragic – but even more eminently celebrated – successors in psychiatry. One of these was "an operation that overtly aimed to damage the frontal lobes of the brain. This operation – *lobotomy* – would win its inventor … a Nobel Prize in medicine [in 1949], … Eventually, it was viewed as the very symbol of psychiatry run amok and eventually abandoned" (p. 284). Prior to this, the advocate of *fever therapy* for "tertiary syphilis and the psychiatric and neurologic catastrophies that accompanied it … received the Nobel Prize in Medicine [in 1927] for his innovation" (p. 280).

6.6 Epilogue

That treatise on "psychiatry run amok" (Section 6.5 above) is, admittedly, an extreme illustration of how medical ethics lost its bearings in the twentieth century. At the root of this particular aberration in medical ethics was the scientific success of Pasteur, Lister, Koch, and others. Inspired by this, *'scientific medicine' got to be the mantra, but the concept of this was severely malformed*; and this fundamentally distorted the medical conceptions of right and wrong, leading to medicine that got to be amoral if not downright immoral (Section 6.1).

In that treatise (ref. 2 in Section 6.5 above), *scientific medicine* is a recurrent theme. In the practice of patient care it meant bringing "the tools and techniques of the new scientific, laboratory-based medicine to bear on the recalcitrant problem of madness" (p. 23). But it also meant, in the Trenton State Hospital, that its "medical staff were encouraged to attend professional meetings, to take professional leaves of absence to further their scientific education, and to undertake original research and publication" (p. 26). The "scientific substance" of the patient care in that hospital was evinced, in a presentation by Cotton himself in a major conference, by "pictures of decaying teeth and charts documenting the presence of harmful bacteria in the gut" (p. 114). 'Scientific medicine' also meant medical science per se, though in a malformed conception of this, as in the statement that the care at the TSH "had accrued from adapting the most recent advances in scientific medicine [meaning medical science] to the treatment of the mentally ill" (p. 175).

With these various conceptions of the essence of scientific medicine (in contrast to the tenable one; Section 4.1.5), "scientific physicians ... insisted that their theories were rooted in the *authority of the microscope and the laboratory*" (p. 36, italics mine). And a physician characterized as a "distinguished authority" officially declared that the work that Cotton described in that major conference "had the *authority of science and real-world results* behind it" (pp. 75–76, italics mine).

The *ethics* of that work was questioned by colleagues; but at issue was not whether it was right "to proceed with dangerous, disfiguring, or even deadly surgery in the face of objections from [the patients]" (p. 97). *The concern was the justification of the care by appeals to lay audience rather than to colleagues.* For, "the medical guild had long railed against actions that imperiled the profession's dignity or that smacked tradesman-like behavior or quackery," trying to "retain the illusion of a professional group that regulates itself" (p. 97). The truth was "the hollowness of the professionals' claims to police themselves" (p. 277).

The author of that treatise – Skull is a sociologist – makes a larger point in this same vein: "We live, as Cotton's contemporaries already did, in an age of experts, professionals who loudly proclaim their disinterestedness, their benevolence, the grounding of their actions in the mental territory that is science. Theirs is a universe of progress, moral quite as much as cognitive and technical. ... Yet in a wide variety of contexts, events of the twentieth century stretched such claims to the breaking point and beyond. Nowhere was that more apparent than when the apostles of rationality confronted the irrational, presuming to divide the mad from the sane and to minister to minds diseased." (p. 278.)

...

So, I may well have been too soft, too benevolent when introducing the two eminent examples of twentieth century medicine as amoral rather than immoral. If Halsted and his followers failed to advise their patients about the relevant truth – that they did not know about the curative implications of their ever more 'radical' operations, that their only concern was thorough extirpation of the tumor at the site of the cancer's inception – then it would have to be concluded that they knowingly failed to act in the best interest of their patients; that is, that their practices actually were immoral. And this was the case in particular insofar as part of the motive was interest in cultivating showmanship in the 'theater' of the operation.

Cotton and his followers and supporters arguably were even more clearly beyond the pale, with their appeals to the authority of science quite possibly plain disingenuous.

As the big-picture paragraph above is a genuine scholar's depiction of modern medicine at large from the ethical perspective, it needs to be weighed and considered with due seriousness, mindful of *the Hippocratic ideal of a doctor free of hubris*, one not only learned but also "wise, modest and humane" (Section 1.1).

Part IV
Pursuing Excellence in Medicine

Chapter 7
Defining Attainable Excellence

Contents

7.0 Abstract

A student of medicine should feel the ethical imperative to develop into a doctor imbued with thorough professionalism – all-around *excellence* – in the services (s)he will provide to clients (Section 6.3). To this end, his/her first-order aim must be the attainment of the maximal possible level of *competence* in some particular discipline of medicine (Section 6.3).

This professional goal needs to be made realistic by defining the discipline in such a way that the level of competence consistent with professional excellence in it actually is attainable. This means the need to define, for oneself, a discipline that is a *pure art* of medicine and *cognitively coherent* as such.

A realistic conception of professional excellence is doing the best anyone could do in the context of the challenges in the discipline and the state of the knowledge-base of coping with them (Section 1.5).

7.1 Defining a Pure Art

The practices described in Section 6.5 fell distinctly short of what the ethical imperative of excellence in medicine (Section 6.3) – of medical *professionalism* in this meaning of it – called for. (The general idea about professionals – in healthcare,

© Springer International Publishing Switzerland 2015
O.S. Miettinen, *Medicine as a Scholarly Field: An Introduction*,
DOI 10.1007/978-3-319-19012-9_7

music, sports, etc. – is that much more is expected of those who are paid for their performances than of those who are mere amateurs at them; It. *amatore*, 'lover.')

In those practices, highly invasive treatments were applied without due regard for treatment-dependent prognoses in diagnosed cases of a particular type of illness (breast cancer); and such treatments were applied without due concern even for diagnoses (and then diagnosis-conditional and treatment-dependent prognoses) in cases of sickness of a particular type (psychosis) and without the patient's consent.

But, were those treatments even intended to reflect professionalism in the practice of medicine and, thus, applied solely in the best interest of the patients (Section 6.3)? or, was the intent in them to learn about the treatments' effectiveness (and safety) by means of experience with the treated patients, perhaps even to demonstrate the effectiveness to colleagues who do not (yet) believe in its existence? In other words, *were those treatments actually experimental* and hence elements in (quite amorphous) medical research rather than in actual medicine, in a semblance of medical science rather than that of scientific medicine (Section 4.1.5)? In both of those examples in Section 6.5 the treatments likely were motivated by the ends of research to some extent at least; and to the extent that this indeed was the case, professionalism of patient care was wanting on this most fundamental level already (Section 6.3).

If a patient with a known case of breast cancer is clearly taught – 'doctored' (Section 2.1) – that even extremely 'radical' mastectomy (in which much more than the affected breast actually is resected) is not known to be more commonly curative than a 'conservative' resection of the tumor together with irradiation of the site of it, and that there isn't even any good reason to entertain the possibility of such a difference in curativeness, (s)he is quite unlikely to submit to the extreme, life-threatening mutilation. But the surgeon's extraprofessional motives are prone to make the 'doctoring' less clear, even unwittingly.

A psychiatrist whose patients are hospitalized for incapacitating psychoses is well aware of how unsatisfactory the various routine treatments for such illnesses actually are, even in alleviation of the symptoms of their underlying somatic anomalies of unknown nature and etiogenesis. As a practitioner of excellent medicine (s)he would simply advise everyone concerned that nothing truly useful can medically be done. But (s)he is tempted not to do the best anyone could do – to acknowledge the impotence of modern medicine in the situation at hand – but to adopt the role of a 'scientist,' however incompetent, in an extreme case à la Henry Cotton and his followers (Section 6.5).

An excellent practitioner of an art of medicine is an all-out professional, thoroughly committed to the provision of the best possible service to his/her clients, with no distraction from or compromise in this from a concern to also help advance the scientific knowledge-base of the practice. Excellence in an art of medicine also means *practice of maximally scientific medicine* – rational in its theoretical framework (Section 4.1.5 and Chapter 5) and true to the state of its scientific knowledge-base, however wanting.

A property of excellent practice of medicine is thoroughness and correctness of its records-keeping for the purposes of the practice. But by such records-keeping, *excellent practice of an art of scientific medicine can also serve, unwittingly, to advance medical science* – by producing data for exploitation by others in their quintessentially applied – gnosis-oriented – medical research (Section 1.4).

7.2 Defining a Coherent Art

With all-around excellence as a practitioner taken to be the overarching imperative of medical ethics (Section 6.3) and indeed the essence of medical professionalism, attainment of thorough professionalism as a doctor is understood to require thorough commitment to serving the client(s), without any compromise in this from service to science (and/or one's personal ambitions) besides (Sections 1.5 and 7.1 above). But there is, also, another, equally important principle in the assurance of the attainability of the highest possible level of professionalism in medicine, having to do with the particulars of the strictly medical focus of one's career in medicine.

In Section 6.5 the focus was on two particular disciplines of medicine, one focused on a particular disease (breast cancer), the other on a particular sickness (psychosis). In both of these the focus may seem to be narrow enough so that the requisite knowledge-base of the practice is cognitively coherent, meaning that expertise in any given aspect of the practice requires mastery of the others, so that the pursuit of a given segment of the expertise is not compromising attention to the others (Section 1.5). But this requires closer examination.

In practices directed to *breast cancer* there are doctors' professional services to the 'worried well,' to address their risks for coming down with an overt case of this disease and dying from it; and setting these prognostic probabilities requires, inter alia, cognizance of the implications of medical ministrations such as prophylactic treatment and the pursuit of latent-stage detection and its associated 'early' treatment of the cancer. Then there are doctors' services to those with a sickness – or laboratory finding(s), radiologic perhaps – possibly due to breast cancer, directed to diagnosis about the presence/absence of this illness and about its differential-diagnostic alternatives in such cases. And for those with a 'rule-in' diagnosis about the cancer, doctors' services center on prognoses conditional, most notably, on staging of the case and various types of treatment, implying the relative effects of the different possible treatments, the differences in their intended and unintended effects.

One notable point about these three types of medical challenge relating to breast cancer is that while two of them are specific to this illness, one is not. The exception is the client presentation with something that could be a manifestation of breast cancer or of any one of its differential-diagnostic alternatives, implied by the generic nature of the sickness or test finding(s). And an additional point of note is that the knowledge-base of sickness-conditional diagnosis about the sickness-explanatory illnesses has no bearing on that of the illness-conditional prognoses.

This example suggests and illustrates a *principle* in the definition of one's art of medicine with a view to attainability of excellence – specifically unsurpassed competence – in it: One is to aim to become a diagnostician or a prognostician, while aiming to be both would tend to mean incomplete mastery of each of them; and if the aim is to be a *diagnostician*, the discipline is to be defined in terms of particular types of sickness, while for excellence as a *prognostician* the discipline is to be defined in terms of focus on particular types of illness. For excellence as an *etiognostician* the focus should be on particular types of either illness or on particular types of sickness not due to illness.

In these terms, cardiology and pulmonology, for example, are not optimally construed disciplines/'specialities' of medicine for diagnostic purposes: When a given type of chief complaint raises the possibility of underlying myocardial infarction or pulmonary embolism (i.a.), the requisite expertise for diagnosis is not intrinsic to cardiology nor to pulmonology; needed is *expertise in differential diagnosis in the context of certain types of case presentation*. And as for the knowledge-base of prognoses conditional on various cardiac illnesses, there generally is little or no cognitive interdependence among them; and the same is true of pulmonary illnesses.

7.3   Defining Competence in the Art

In the Hippocratic Oath the newly-minted doctor swears to function "according to my ability and judgement" (Section 6.4); but his/her ability and judgement may be quite wanting, well short of the excellence that is the overarching ethical obligation in medicine (Section 6.3).

While an introduction to medicine is not the context for elaboration of all that generally goes into excellence in a doctor's practice of an art of medicine, it is the context to define the aimed-for level of excellence in suitable relation to the objectively-defined ideal. For this I suggest the following *principle*: A doctor-to-be is to so define his/her future art of medicine that in all of the tasks involved in it (s)he will be able to achieve a level of competence unsurpassed by (practically) anyone; (s)he is to be clear on the boundaries of the domain of his/her aimed-for excellence and be determined not to routinely take on tasks for which (s)he is less competent than some other doctors.

This is to say that *competence in medicine* is a relative concept: it means ability to function on the level that characterizes the best in the particular task (gnostic or other) that is at issue. This level of competence in the task at issue should not be viewed as being an exception; it should be held as the *norm*. For only in this way can all of the routines in medicine, across all of the practitioners who engage in them, be excellent. By suitable definitions of their disciplines, doctors can assure this; and their clients have the right to expect this.

The aspect of excellence at issue here is *full competence*; and this generally (though not always) means excellence in the *gnoses* of concern in the discipline, not decision-making (Section 5.7).

Chapter 8
Pursuing Attainable Excellence

Contents

8.0 Abstract

Board certification of qualifications in a given discipline of medicine does not mean that one is ready for competent practice in that discipline; the prevailing general idea is that full competence as a diagnostician, most notably, is the result of subsequent extensive *experience* with practice. This means that the greatest attainable competence in a doctor's professional work actually is seen to be attained only late in the career, and only if (s)he by then indeed has gained extensive experience and has optimally learned from it.

A student of medicine heeding the ethical obligation to develop into an excellent doctor not only defines his/her future discipline with a view to this (Sections 7.1 and 7.2); (s)he also adopts a plan for maximizing the rate of learning from experience in that discipline.

But before this (s)he *needs to learn how challenging it is to learn about diagnostic and other gnostic probabilities on the basis of case-by-case experience*, and that the source of genuine knowledge about gnostic probabilities ultimately is not practitioners' case-based learning but scientists' *gnostic research*.

© Springer International Publishing Switzerland 2015   111
O.S. Miettinen, *Medicine as a Scholarly Field: An Introduction*,
DOI 10.1007/978-3-319-19012-9_8

8.1   Being Realistic About Case-Based Learning

Let us consider, again, the case-based/'problem-based' learning that begins in medical school (Section 2.4) and continues upon graduation in a residency program and then throughout one's career as a doctor; and let us focus on learning to set *diagnostic probabilities* on the basis of experience with cases of a given type of sickness presentation for diagnosis.

At issue here is learning to set the various profile-specific diagnostic probabilities for the presence of the particular illnesses in the differential-diagnostic set implied by the type of case presentation (chief complaint, demographic category; Sections 4.2.2 and 5.2). Of course, instructive about those probabilities only are cases in which the truth about the presence/absence of each of the illnesses in that set became known on the basis of subsequently emerging facts (completing a pathognomonic diagnostic profile), and all of the relevant facts (in the diagnostic profile) became known, specifically, to the doctor in question. The cognitive challenges in this are huge, as outlined in Section 5.5. That introduction to the topic, quite discouraging, is made more concrete here.

The focus here is on a very *simple hypothetical example*, involving only three diagnostic indicators (defining subdomains of the presentation domain), two of these of the binary ($+/-$) type, the third quantitative. For the first 40 informative cases one encounters, the relevant facts could be those in the attached Table 8.1, including the datum on the subsequently-established presence/absence of the illness at issue. Also

Table 8.1 Experience as a teacher of diagnostic probabilities

D_1	D_2	D_3	I	D_1	D_2	D_3	I	D_1	D_2	D_3	I	D_1	D_2	D_3	I
+	+	125	+	+	+	134	+	−	−	93	−	+	+	111	+
+	−	91	−	−	+	127	+	−	+	124	+	−	−	75	−
+	−	111	−	+	+	105	+	+	−	110	−	+	−	100	−
−	−	99	−	+	+	127	+	+	+	124	−	+	+	124	+
+	+	114	+	−	−	87	−	−	−	108	+	−	−	99	−
+	+	131	+	−	−	108	+	−	+	103	−	+	−	86	−
+	−	94	−	+	+	139	+	+	+	125	+	+	+	121	+
+	+	120	+	−	−	98	−	+	+	118	−	+	−	98	−
+	+	110	+	−	+	112	+	+	+	114	+	+	+	117	+
−	−	104	+	−	+	117	+	+	−	82	−	−	−	99	−

Based on this experience, thus documented and layed out, the ED doctor is to set Pr(I) in subsequent cases from the same domain conditionally on (D_1, D_2, D_3) of $(-, -, 100)$ and $(+, +, 120)$, etc. See text.

Situation (hypothetical): Given a particular presentation (chief complaint, demographic category) of a patient in hospital's emergency department, diagnosis about illness I is first considered upon ascertainment of three additional items of information, and further action depends on the probability of the presence of illness I (i.a.). A doctor in the ED is familiar with 40 cases with the presentation at issue, such that the truth about the presence/absence of I became known (after the initial diagnosis). For these 40 cases, (s)he has layed out the data additional to the type of presentation – that is, D_1, D_2, D_3, and the I status – as shown here shown in the table are two subsequent cases (also hypothetical) for application of the learning from those 40 cases, for setting the diagnostic probabilities for the illness at issue in those cases.

Given these data, a very instructive exercise for the reader would be a serious attempt to learn, informally, just as doctors generally aim to do, the various profile-specific probabilities for the presence of the illness in question; that is, a serious attempt to acquire such 'tacit knowledge' manifest in its application to subsequent cases, such as the two specified in that Table 8.1, in being able to set the respective diagnostic probabilities in these two cases.

The learning that should be contemplated upon the completion of this exercise has to do with answers to these questions: How instructive is that experience with the 40 cases – fully documented and expressly layed out! – about the probabilities in those two cases? Could different reviewers of these same data arrive at appreciably different probabilities in either one of those two cases? And: Could the information content be very different in the ensuing, new set of 40 informative cases from the presentation domain at issue?

The learning from this exercise – and its consequent realism in regard to case-based learning about diagnostic (and other gnostic) probabilities – can be consolidated by studying Appendix 2. For, the implications of those 40 cases to a diagnostic researcher are formally addressed there, including the degree of reproducibility of the probability results in their application to those two cases, all of this supplemented by delineation of how the experience can be misleading on account of invalid entries into the set of available informative cases.

While this exercise on case-based learning about diagnostic probabilities in the practice of medicine is very discouraging about the feasibility of this learning in the context of the simple hypothetical example herein addressed, its implications are even more discouraging in respect to the 'tacit knowledge' about diagnostic probabilities that supposedly is gained in actual practice. For the real experiences commonly involve quite complex specifications of the case profiles, fading recollections of the facts in the instances in which the truth became known, and little or no attention to potential major biases in the 'self-selection' of cases into the experience that is the basis for the learning.

More on this in Appendix 2.

8.2 Being Realistic About Science-Based Learning

Section 8.1 above supplementing Section 5.5 – especially if both of these are supplemented by Appendix 2 – should give a student of medicine serious second thoughts about something (s)he may have enthusiastically adopted from Section 1.1: the Hippocratic philosophy that a physician learns to practice medicine through observations on the cases (s)he encounters and reasoning about these. Recall the failure to learn anything about the effects of the ubiquitous bloodletting through the millennia of Hippocratic-Galenic medicine, extending to the twentieth century (Section 1.3).

The main challenges in the practice of *modern* medicine unquestionably are diagnostic and intervention-conditional prognostic *probability-settings*, bearing on knowledge-based decisions about the choice among the possible actions (gnostic or interventive); and alas, *these probability-settings a doctor's personal, case-by-case experience does not meaningfully teach*, however well this experience is documented and whatever powers of informal reasoning (s)he may bring to bear on these facts in an effort to learn the probabilities from it (Section 8.1 above).

. . .

So, what about *science* as a teacher of gnostic probabilities for the practice of medicine? a thoughtful student preparing for a career in modern medicine asks.

The first thing to learn about this post-Hippocratic, modern possibility is that, contrary to what now is commonly taught (Sections 2.4.1 and 4.1.1), *modern medicine itself is not science* (Section 4.1.1). Nor does it deploy 'the scientific method' (hypothetico-deductive or whatever be the concept of this) in arriving at gnostic probabilities (prognostic ones in particular).

Second: While in an avant-garde medical education "it is made profoundly clear that learning medicine in the first 2 years [in a 4-year medical education toward the MD degree] is above all learning the biomedical sciences" concerned with "cell biology and physiological systems," inter alia (Section 2.4.1), the student needs to be realistic about the relevance of this: The contents of these studies may be seductively interesting, and they may provide for understanding of some phenomena of illness; but medicine does not depend on interests and understandings. *The need in medicine is for gnosis-serving knowledge, and those 'biomedical' sciences – the 'basic' sciences of medicine – do not provide any semblance, even, of the knowledge-base for setting gnostic probabilities.*

And third: This realism about the 'basic' medical sciences as a source of the knowledge-base for medicine should not lead to the misunderstanding that scientific medicine – characterized by a rational theoretical framework together with knowledge-base from science (Section 4.1.5) – is not realistic to dream about. Rather, the need is to understand that *the scientific source of the requisite knowledge is medical science in a meaning very different from 'biomedical' science or 'basic' medical science*, which is only one of the principal lines of medical science (Section 1.4).

The source of the scientific knowledge-base of medicine is what I think of as *quintessentially applied medical research* (Section 1.4). It is *statistical science* for medicine. An introduction to this I provide in my most recent previous book (*Toward Scientific Medicine*). Suffice it here to merely note that such research for clinical medicine, addressing gnostic probability functions, has not yet even begun, at least not in earnest (cf. Section 1.4.2 and Appendix 1).

8.3 Practicing the Art of the Possible

Having come to appreciate these orientational truths about the knowledge-base of the 'learned' professions of medicine, a student starting to prepare for a career in one of the disciplines of modern medicine inescapably is quite frustrated: Even with focus on medicine proper, without distractions from dreams of also contributing to medical science (Section 7.1), and with one's discipline of medicine defined with a view to cohesiveness of its requisite knowledge-base (Section 7.2), the happiness attendant to professional excellence (Section 6.3) – in the ultimate meaning of humdrum pursuit and attainment of appropriate gnostic probabilities (Appendix 3) – remains unattainable at present and also in the near-term future.

A suitably wise student of modern medicine, however, does not dwell on this frustration. (S)he sets out to practice, as a medical student already, the art of the possible – and this with focus on what actually is learnable from and also relevant to practice in the discipline of his/her choice (which may not be officially recognized as a 'specialty' or, even, as a 'subspecialty'). And actual learning of the learnable-and-relevant (s)he really needs to pursue, as a suitable practice-guiding 'expert system' (incorporating the relevant knowledge) remains unavailable for whichever discipline of modern medicine. (A dream of the future of knowledge-based practice I outline in Appendix 3).

In the pursuit of the possible-and-relevant learning for a particular discipline of modern medicine, the student needs to have a suitably-ordered specification of the agenda, tailored to that discipline.

Insofar as *sickness-prompted pursuit of diagnosis* is a/the concern in the discipline, possible and relevant to learn – and to make a record of – are the types of case presentation that my occur (ordering these by their respective commonalities). Distinctions among them are to be made if, and only if, they have bearing on the process that ensues.

Separately for each of these distinct types of presentation, possible and relevant to learn is, first, what set of facts are to be ascertained (routinely) for the first-stage, clinical-diagnostic profile, and what the complete set of differential-diagnostic possibilities is – and to again make a readily accessible record of this (as though content in the discipline-specific 'expert system').

Then comes the first truly difficult part: learning to set the probabilities for each of the differential-diagnostic possibilities, separately for each of the various diagnostic profiles that are possible in the context of a given type of case presentation. To this end it would be necessary to document informative cases analogously to Table 8.1. However, informal learning of these probabilities is prone to be next to impossible (Sections 8.1 and 8.2 above); but at least learnable is what it is that remains unknown – and this can be put to use in terms of advocacy of the relevant research and, also, supply of data for it (from practice-serving record-keeping; Section 7.1).

. . .

This sketch of the learning agenda for a very characteristic type of gnostic challenge in medicine (within a given discipline of it in modern medicine) might serve as a paradigm for developing its counterparts for other generic types of gnostic challenge. But the larger point here is the one above, namely that, with one's discipline of medicine suitably defined, Hippocratic-type learning toward excellence in it (Section 1.1) is possible, but to an extent only (while knowledge from the relevant, quintessentially applied research remains but a dream).

The neo-Hippocratic practitioner of modern medicine, imbued with wisdom and modesty as (s)he is (Section 1.1), does not dwell on and be unduly frustrated by the insurmountable gnostic challenges of practice; (s)he accepts the most eminent one of the Hippocratic aphorisms (cited in Section 1.1 already): "Life is short, the art long; timing is exact, experience treacherous, judgement difficult," notably as to how wanting experience is as a teacher of gnostic probabilities.

With *excellence* the mantra in all of this (Section 6.3) and competence a requirement for this, I suggest that the concept of this quality of a doctor's care be taken to be one constrained by realism: in respect to any given gnostic challenge, competent gnosis is *the best gnosis anyone can suggest*, given the prevailing state of the general knowledge-base for the gnosis at issue (Section 7.3).

Part V
Epilogue

Chapter 9
The Meaning of It All

Contents

9.0 Abstract

A particularly illustrious professor of mine in medical school (in the 1950s) subsequently remarked to me, privately, that in medical academia "nothing is new, nothing is true, and nothing really matters." As a counterpoint of sorts to this, I now say that the propositions in this book are largely new to medical academics, and that the respective truths do matter (cf. Preface). But it remains to the reader to weigh and consider to what extent (s)he can accept those propositions as representing truths (in the theory of medicine), and to what extent (s)he can agree that those truths, whatever they may be, really matter.

9.1 Gnostic Knowledge as the Key

A particularly illustrious professor of mine in medical school (in the 1950s) subsequently remarked to me, privately, that in medical academia "nothing is new, nothing is true, and nothing really matters." As a counterpoint of sorts to this, I now say that the propositions in this book are largely new to medical academics, and that the respective truths do matter (cf. Preface). But it remains to the reader to weigh and consider to what extent (s)he can accept those propositions as representing truths

© Springer International Publishing Switzerland 2015
O.S. Miettinen, *Medicine as a Scholarly Field: An Introduction*,
DOI 10.1007/978-3-319-19012-9_9

(in the theory of medicine), and to what extent (s)he can agree that those truths, whatever they may be, really matter.

Having reached the terminus of the academic journey through this book, the reader might recall the beginning of it. In the Preface I laid out *eight questions* as examples of what a suitable introduction to medicine would help the beginning student of the field to think critically about. Now the reader might pause and think, critically, about his/her own answers to those questions. And having done this, (s)he presumably would be interested in my corresponding answers, which I give in Appendix 4. As for the differences between these two sets of answers, it would be good to reflect on whether a discussion of these differences likely would lead to agreement on a single set of answers. This juxtaposition of views would give a little taste of the now-absent discourse among medical scholars that, in the fullness of time, would establish a canonical body of writings on the general theory – philosophical – of medicine (Section 2.2).

In the introductory matters of medicine, I suggest, it would already be progress if there were to be agreement, and teaching, that absence of health means presence of *illness*, commonly a hidden somatic anomaly at least potentially manifest in sickness; that *diagnosis* has to do with the presence/absence of a particular type of such (hidden) illness and is probabilistic knowing (esoteric) about its presence/absence (rather than "the determination of the nature of a case of disease" or "the art of distinguishing one disease from another"; Section 4.2.1).

As the prime example of the needed progress in medical academia along those lines – in the advancement of medicine as scholarly field – I am proposing an express conception of what *medicine* is, namely that the essence of medicine is *gnostic knowing*, founded on *gnostic knowledge* (Section 4.1.1). And of considerable note in this context is the conceptualization of *surgery* from this vantage: Is surgery to be seen as a field lateral to medicine, as is implied by the common expression 'medicine and [*sic*] surgery'? Or is it, as my dictionaries of medicine say, a "branch of medicine"? Or is it, as I suggest, best construed as a particular modality of treatment within medicine, along with radiotherapy and pharmacotherapy, for example? (Section 4.1.2.) Insofar as a discipline of 'surgery' actually is one of medicine, in its essence I take to be gnosis and merely incidental to it I take to be the feature that the doctor may carry out a surgical intervention (opted for by the patient in the light of intervention-prognoses – 'conservative' interventions included – taught by the 'surgeon').

Directly related to this, I note that in Canada the education and training in "specialties" of "medicine" – "specialty medicine" – is in the policy domain of an agency lateral to (or above?) medical academia: *The Royal College of Physicians and Surgeons*. It has, notably, a Division of Medicine and separate from this a Division of Surgery. By its name this august agency clearly implies that Canadian surgeons – with their required MD degrees and licenses to practice medicine – are not physicians. This I take to be something that medical academics, in Canada at least, need to seriously weigh and consider. For if a surgeon is merely someone who carries out operations opted for by doctors' patients, (s)he is not a doctor. But if a so-called surgeon actually pursues, achieves, and teaches (to patients) intervention-prognosis in all relevant respects, (s)he is a doctor – on this basis rather than that incidental competence to carry out operations.

The reader will have noted that while I don't regard surgery as a field distinct from medicine, I make a sharp distinction between medicine and medical science, and that I underscore the deployment (and not the development) of *general medical knowledge* as being in the essence of medicine, central to doctors' pursuit and (potential) attainment of esoteric knowing – gnosis (dia-, etio-, and/or pro-) – about the health of their clients. Related to this, the reader will have noted how I lament the prevailing state of medical science in the production of the knowledge-base for medicine, for gnosis in it (Sections 1.4.1 and 1.4.2, i.a.).

Another major theme in this essay, the reader will have noted, is the doctors' ethical obligation to be thoroughly professional in their services to their clients, the essence of which I take to be all-around, practically unsurpassed excellence in those services. And in sketching the avenue to the excellence, I underscored *what the excellence is first and foremost about, namely gnoses per se* rather than (gnoses-based) decisions about actions (gnosis-serving or interventive), and that *guidelines for practice also should focus on gnoses* rather than actions (Section 5.7).

9.2   Gnostic Knowledge as the Goal

Regarding gnostic excellence in medicine at present, eminent doctors are prone to take pride and joy in their 'studies'/'investigations' on the health of their patients, in the creativeness and other virtues of their cognitive processes in medical 'problem-solving,' and in their 'scientific' preparedness for this. And as for gnostic probability-setting, specifically, they tend to think of themselves as experts in the meaning of being better doctors than many of their colleagues in the same discipline, and the main basis (much overrated; Section 8.1) for this tends to be their relatively extensive personal experience in practice.

To them it commonly is an anathema that the knowledge fundamental to gnostic excellence might be codified, for deployment by any doctor, in computerized algorithms for fact-finding on the clients and for humdrum transitions from the facts to expertly gnoses – in practice-guiding *'expert systems'* (Section 5.6 and Appendix 3). They tend to disparage practice of medicine under the guidance of 'expert systems' – if and when they become available – as something for 'technicians' rather than worthy of true doctors. They tend to have an aversion to visions of *expressly codified gnostic knowledge as the central goal in the advancement of medicine*, preferring an air of mystique about their gnostic 'tacit knowledge.'

When senior doctors with such disdain for 'expert systems' book a commercial flight, they do not check the credentials of the captain in charge of the flight: they trust that any professional in commercial aviation is educated and computer-guided to act the same way in any given situation – the way that objectively is in the best interest of the clients these professionals serve. Senior doctors are happy to be able to rely on such professionals. And I am unaware of modern airline captains lamenting that they in this Information Age have been made into mere technicians, pining for the time when they were creative problem-solvers.

Only if guided by genuinely knowledge-based gnostic 'expert systems' would medicine truly be knowledge-based. But the availability of the requisite knowledge remains but a dream (Appendix 3), as industry lacks the profit motive for its development and medical academia lacks understanding of the requisite theoretical framework for this (Section 1.4.2 and Appendix 1).

9.3   Philosophy as the Guide

In closing here, it needs to be noted that *education for the learned professions has been the express concern of the Carnegie Foundation for the Advancement of Teaching* ever since its founding in 1905 (and chartering by the U.S. Congress in 1906). And so, this present work of mine addresses introduction to one type – the preeminent one – of the professions in this august institution's purview.

Under the auspices of this Foundation, as I've already noted, was developed the highly influential '*Flexner report*' on medical education, published in 1910 (ref. 2 in Section 4.1.5). And recently, also under this Foundation's auspices, an *update of that report* (ref. below) was published.

Reference: Cook M, Irby DM, O'Brien BC (Foreword by L. S. Shulman). *Education of Physicians. A Call for Reform of Medical School and Residency*. San Francisco: Jossey-Bass. A Wiley Imprint, 2010.

This new report points out that "The basic features outlined by Flexner [for medical education in the U.S. and Canada] remain in place today: a university-based education consisting of two years of basic sciences and two years of clinical experience in a teaching hospital" (p. 1). And in the required 'pre-med' studies (for a Bachelor's degree, in the U.S. and Canada), "The standard courses are a Flexnerian legacy and include one year of biology, two years of general and organic chemistry, one year of physics, and in some schools one year of mathematics" (p. 19).

The *fragmentation of medicine* into its constituent disciplines that has been so eminent an element in the evolution of medicine since the Flexner report – and is the point of departure in my Preface here – is noted in the Foreword of this new report, by the Foundation's President Emeritus. But the educational implication of this evolution he sees to be very different from what I posited in the Preface here. He says that "medical students are expected to learn all these domains and somehow to connect, combine, and integrate them with their own understandings and their own professional identities" (p. ix).

In that report, nothing is said about the principal premise in this present oeuvre of mine, namely the *need to introduce the students to the newly-complex field of medicine at large* (Preface). Said is, instead, this: "The first phase of medical school should establish the patterns that physicians will use throughout their lives to link their clinical work to continuing learning. Specifically, it must introduce students to the dialectical relationships that establish 'the spiral learning helix,' ... " (pp. 219–220).

The authors do, however, also address *the very beginning of medical-school education*. They say that "our design has medical school beginning with an omnibus course," continuing with the specifics of it: "This course consists of four elements: (1) advances in molecular biology and genetics, (2) foundational social sciences for medicine (anthropology, psychology, and sociology), (3) biostatistics and introductory epidemiology, and (4) medical humanities. Students are expected to take the two elements in which they did the least course work as undergraduates or in which they showed the least proficiency on placement testing" (pp. 220–221).

Regarding this Flexner-redux conception of the proper nature of the opening of education in modern medicine in juxtaposition to the content I have here put forward, I repeat, yet again, the counsel by Francis Bacon I quoted in the Preface already: "Read not to contradict, nor to believe, but to weigh and consider."

As the contrast is very stark, it thereby well illustrates a very important point: **philosophy matters**! Modern medicine was founded on philosophy (Section 1.1), and much progress was the consequence of this (Section 1.3). As philosophy is the mother of science, the scientific progress in medicine (Section 1.4), too, has its foundation in philosophy. And now, as that contrast well illustrates, philosophy has a profound bearing on how students of medicine are introduced to this field of scholarship and, thereby, on what is seen to be essential in the development of competence as a scholar in whichever discipline of modern medicine.

The crucial role of philosophy (explicit or implicit) in shaping the introduction of students of medicine to this field – once a module devoted to this becomes an element in the medical common (Preface) – means that their teachers, senior academic physicians among these in particular, need to be suitably imbued with the relevant philosophy – *critical philosophy of medicine*, that is. Aristotle, already, called for this, as "One of his precepts was that the philosopher must begin with medicine and the physician must end with philosophy" (ref. in Section 1.1; p. 70).

Appendices

Appendix 1: Theoretical Framework for Diagnosis

Scientific medicine, it may be recalled from Section 4.1.5, is characterized by a *rational theoretical framework* together with deployment, in such a framework, of substantive knowledge derived from science. The framework and knowledge pertain to diagnosis, etiognosis, and prognosis (Sections 5.2, 5.3, and 5.4). The theoretical framework of gnosis (dia-, etio-, pro-) should be rational even if the type of knowledge that its application calls for is not available. Illustrative of all of this is the theoretical framework of diagnosis.

The theoretical framework for diagnosis that still is commonly taken to be rational – by 'clinical epidemiologists,' most notably – was adduced over a half-century ago by a dentist together with a radiologist (Ledley & Lusted, *Science*, 1959). The basic idea was that the probability of the presence/absence of a particular type of illness is not knowable directly when known is the diagnostic profile of the case; that knowable, even in principle, only are the *reverse probabilities*, the probabilities of the manifestations profile of the case in the presence and absence of the illness in question; and that, therefore, the diagnostic probability at issue needs to be derived by means of *Bayes' theorem*: Let $Y = 1$ and $Y = 0$ represent the presence and absence of the illness in question, respectively; and let X represent the vector of statistical variates that were adopted to represent the manifestations-based diagnostic indicators, with $X = x$ this segment of the diagnostic profile of the case at issue. Bayes' theorem gives the 'posterior,' x-conditional probability of the presence of the illness, P'', as a function of the 'prior,' manifestationally unconditional counterpart of this, P', together with the likelihood ratio, LR, for the (vector-valued) datum, $X = x$:

$$P'' / (1 - P'') = [P' / (1 - P')] \times LR;$$
$$LR = \Pr(X = x \mid Y = 1) / \Pr(X = x \mid Y = 0).$$

© Springer International Publishing Switzerland 2015
O.S. Miettinen, *Medicine as a Scholarly Field: An Introduction*,
DOI 10.1007/978-3-319-19012-9

This theoretical framework for diagnosis is logically *untenable*. Bayes' theorem is correct, of course. But the LR is, in the theorem's application here, a malformed concept; and even if it weren't, diagnostic knowledge of that reverse-probability (or probability-density) form would represent generally unsurmountable epistemic challenges.

Any LR is the ratio of two conditional probabilities (or probability densities), and conditional probability is meaningfully defined only if the condition is singular in its meaning, in its bearing on the probability at issue, here $\Pr (X = x)$. Illness, when present at the time of diagnosis, is generally a highly nonsingular type of entity as a cause of its manifestations. And when the illness at issue is not present to explain the diagnostic profile, its alternative is by no means singular either. The LR in this formulation is, thus, generally devoid of express meaning, even in principle.

Let us nevertheless assume, counterfactually, that illnesses targeted for diagnosis generally are singular entities as explanations of the illness-manifestational profile, so that the LR in that Bayes' theorem formulation of diagnostic probability actually is a defined, single-valued quantity. Now the need is to appreciate that the various possible values of the LR are not subject to becoming known as to their magnitudes. Suffice it to think of $\Pr (X = x \mid Y = 1)$. Valid study of this requires access to cases of $Y = 1$ (i.e., of cases of the illness at issue) independently of the manifestations profiles $X = x$. This is, however, generally unimaginable. And if this weren't a problem, the enormous number of the possible profiles conditionally on $Y = 1$ would generally make study of their respective probabilities wholly impracticable.

Despite its untenability, that theoretical framework for diagnosis, centering on Bayes' theorem, has been adopted with great enthusiasm by 'clinical epidemiologists.' Central in this naturally has been the particular topic of the updating of a '*pretest*' diagnostic odds, based on the pretest diagnostic probability, P′, by a test result to the corresponding '*post-test*' diagnostic odds, involving the post-test probability, P″:

$$P'' / \left(1 - P'' \right) = \left[P' / \left(1 - P' \right) \right] \Pr (T = t \mid I) / \Pr \left(T = t \mid \bar{I} \right),$$

where $T = t$ denotes the result (t) of the test (T) at issue, and where I and $\bar{I}$ denote the presence and absence, respectively, of the illness at issue.

Even though diagnostic tests generally produce quantitative results, in this particular context there is an eminent tradition to reduce the quantitative result to a binary one, either 'positive,' T+, or 'negative,' T-, which point, respectively, to the presence and absence of the illness in question. And a well-established terminology is in use to denote the probabilities of these binary results:

$$\text{Sensitivity}: \Pr (T + \mid I),$$
$$\text{Specificity}: \Pr \left(T - \mid \bar{I} \right).$$

In these terms,

$$\text{LR for T+} : \text{Se} / (1 - \text{Sp}),$$
$$\text{LR for T-} : (1 - \text{Se}) / \text{Sp}.$$

These ideas have been extended beyond actual test results in the diagnostic profile in the manifestational profile, and even to the risk indicators in the full profile, in terms of decomposition of the overall LR to the 'test'-specific LR_1, LR_2, etc.:

$$\text{LR} = LR_1 \times LR_2 \times LR_3 \times \ldots,$$

thus overcoming the serious epistemic problem alluded to above. This, in turn, has led to a *huge industry* of studies on the 'sensitivity' and 'specificity' of various 'tests,' and also of 'meta-analyses' of their results. In some of these original studies, the 'test' results are addressed in terms of more detailed, ordinal categories, and LR values specific to these are then addressed.

Textbooks of medicine, and practitioners similarly, have taken but token notice of these ideas and of the products of the industry based on these, and this is just as well. Linguistically, it makes little sense to say that taking the history of myocardial infarction is a 'sensitive test' for the presence of a fresh myocardial infarction, or that electrocardiography as to ST level is a 'specific test' for the presence of pneumonia, even though both of these ideas are consistent with that inclusive definition of diagnostic test and the two performance properties of it. And conceptually, the problems with 'sensitivity' and 'specificity' are even more serious than those with the Y-conditional probabilities of $X = x$ above. The added problem now is that they, and each LR_i based on them, should be conditional on the set of profile elements accounted for before the ith 'test.'

The genuine concept of *sensitivity* of a genuine diagnostic test would be the tendency of its result to change from what it would be (or was) in the absence of the illness to what it would be (or was) in the presence of the illness; that the test result tends to reflect the actual presence/absence of the illness at issue (as the test tends to 'sense' the presence of the illness) rather than, notably, reflect the risk to contract the illness. Thus, in the diagnosis about myocardial infarction, history of MI has no true sensitivity for a current, fresh MI, even though it is an important one among the diagnostic indicators, nor is the taking of this history a diagnostic test (contrary to the teachings of 'clinical epidemiologists'). And genetic testing (when the illness is not genetically defined) has no true sensitivity to the illness, however valuable the genetic risk information may be in the diagnosis about the illness.

A test's *specificity* to a particular illness cannot be given any tenable definition; but a particular *result* of the test can be specific to the *presence* of the illness, meaning that this result occurs only in the presence of the illness, that it is pathognomonic about the presence of the illness; and a result of a test can be specific to the *absence* of the illness, pathognomic about this. Examples of specific results of diagnostic tests include the specificity of a positive result of the troponin test to the presence of (a fresh case of) myocardial infarction, and the specificity of a negative/normal result of the chest-radiographic test to the absence of pneumonia.

. . .

A *tenable* theoretical framework for diagnosis is implicit in the nature of the object of it. Diagnosis – a form of knowing – is pursued in a particular generic *domain* of this (e.g., in the domain of the adult with chest pain as the chief complaint; Section 5.2); and the object of diagnostic knowing – probabilistic – about the presence/absence of illness I is:

> Pr (I present, given the diagnostic profile of a case from the domain) ;
> Pr (I | X = x) ,

where $X = x$ denotes the *entire* diagnostic profile of the case.

In statistical terms, this is the probability, P, that a random variate, Y, indicating the presence/absence of I (Y = 1 if I present, 0 otherwise) takes on the realization Y = 1 conditionally on the realization $X = x$ of a set of variates based on the diagnostic indicators involved in the (entire) diagnostic profile of the case. And a common type of statistical model for this type of situation is the *logistic* one:

$$\text{Log}\,[P/\,(1 - P)] = B_0 + \Sigma_i B_i X_i.$$

In this theoretical framework, once the Xs are defined, the object of the requisite knowledge is the set of values of the parameters (B_0, etc.) involved in the adopted logistic model.

Different from the reverse probabilities involved in the theoretical framework of diagnosis centering on Bayes' theorem (above), this logistic framework is one of well-defined probabilities whose values are subject to study – by identification of a series of instances of the presentation domain, documenting Y and X, and fitting the model to the data. Diagnostic research in this rational theoretical framework is a prerequisite for scientific diagnosis; but it has barely begun, while research in the false framework of addressing diagnostic indicators' 'accuracy' – 'sensitivity' and 'specificity' – for particular illnesses continues to flourish (cf. above).

. . .

In the continuing absence of meaningful diagnostic research, diagnostic probabilities are set on the basis of *practitioners'* 'tacit knowledge,' derived informally from personal experience with cases of various diagnostic profiles from particular domains of case presentation.

Formally, the idea in this is that once a diagnostician has had experience with cases from a given domain – cases of the two possible realizations of a Bernoulli-distributed Y in association with various realizations of vector X of profile variates – (s)he has thereby learned how to set Pr (Y = 1 | X = x) for a new case from that domain of case presentation – informally, without the statistical framework sketched above and without, even, any thoughtful review of the cases.

This idea – that a diagnostician's personal experience with cases in which the truth about the presence/absence of a particular illness gets to be known teaches him/her how to set diagnostic probabilities for the presence of the illness – is by no means easy to justify on a-priori grounds (Section 8.1); and empirically, 'expert' diagnosticians' diagnostic probabilities are known to be quite discordant in the context of particular instances of diagnostic challenge. But the idea endures that personal experience with cases is the basis for diagnostic expertise – while textbooks of medicine don't teach the knowledge-base of diagnosis, and education in whichever discipline of medicine doesn't.

This idea of practice-based learning of diagnosis therefore should be subjected to *research in cognitive psychology*, which likely would show that the idea is a myth (Section 8.1), at great variance with the truths about the learning potential of the human mind in this particular, very challenging context (outlined above).

In this psychological research, the focus could be on a very simple logistic model, with only two diagnostic-indicator variates, one binary and the other quantitative. For select realizations of these variates (X_1 and X_2) in N instances ('cases'), the corresponding chance realizations of Y (0 or 1) would be produced under the adopted model. The N instances of realizations of the three variates (X_1, X_2, and Y) would be presented to the study subjects, for them to learn how $\Pr(Y = 1 \mid X_1 = x_1, X_2 = x_2)$ could be set in some new instances from the presentation domain, with known realizations of the Xs but the respective realizations of Y unknown. Then, some new instances of the case profile would be presented to the study subjects, and they would be asked to set the corresponding probabilities for Y = 1, based on the experience with N 'learning' cases. The variates and their realizations would be presented in realistic substantive terms, and the participants would be arranged to have a motive to learn how to set $\Pr(Y = 1 \mid X = x)$.

If the idea of diagnostic learning from personal experience would be shown to be untenable by such psychological research, this would provide a strong impetus for medical research on diagnostic probabilities – under well-designed logistic models. See Appendix 2.

The theory of diagnosis and diagnostic research I address more extensively in my most recent previous book (*Toward Scientific Medicine*. New York: Springer, 2014), along with the counterparts of these for etiognosis and prognosis.

Appendix 2: Researcher-Assisted Learning from Practice

Section 8.1 was an introduction to realism about practice experience as a basis of learning about diagnostic probabilities. The focus was on a very simple example (hypothetical) of experience relevant to this learning; and the point of it was illustration of a grim fact: informally experience-based 'tacit knowledge' about

Table A2 Experience in Table 8.1 of Section 8.1 replicated, with the same size

D_1	D_2	D_3	I	D_1	D_2	D_3	I	D_1	D_2	D_3	I	D_1	D_2	D_3	I
−	+	114	−	+	−	116	+	+	+	119	+	+	+	122	+
−	−	99	−	+	−	94	−	−	+	109	−	−	−	79	−
+	+	130	+	−	−	103	+	+	−	105	+	−	−	102	−
+	+	108	+	−	−	104	−	−	−	84	−	−	−	102	−
+	−	109	−	+	−	110	−	+	+	119	+	+	+	123	+
+	+	122	+	+	+	133	+	−	−	113	−	+	−	110	+
−	−	89	−	−	−	105	−	−	−	101	−	+	+	96	+
−	−	115	−	−	−	106	+	+	+	116	−	+	+	110	+
−	+	127	+	−	+	117	+	−	+	110	+	−	−	109	−
+	+	127	+	−	−	105	−	+	+	112	+	+	−	96	−

See text

diagnostic probabilities is prone to be very insecure, perhaps not even qualifying as knowledge (Section 4.5.1). And in Section 8.2 the related main point was that practice experience can be truly instructive to the practitioner only through the intermediary agency of the diagnostic researcher.

For understanding and acceptance of this principle of case-based ('problem-based') learning of diagnosis, to focus on this pivotal species of gnosis, the student of medicine may need to know, in broad outline, how the researcher thinks about the problem and how (s)he in this theoretical framework uses the data to derive estimates of the probabilities at issue.

The researcher, if suitably oriented, thinks about the diagnostic probability for the illness in question as a function of the set of diagnostic indicators in the domain of the case presentation (Appendix 1). Having designed the form of the function (s)he fits it to the data. And the thus-obtained empirical probability function (s)he provides to the practitioner concerned to learn from the cases (s)he knows and has documented for statistical 'analysis' (synthesis) – and others – for reading the profile-specific probability estimates as they are needed in practice.

As for the example in Section 8.1, a suitable design (below) of the probability function's form and its fitting to the data in the table (Table 8.1) there leads to objective estimates of the probabilities in the two cases specified on the bottom of the table. They are, respectively, 0.15 and 0.85.

But a note of *great caution* is in order about the probability function based on those data. For one, the data could just as well have been the ones in Table A2 here; and based on these data, the corresponding estimates of those two probabilities are quite different from the ones above: 0.07 and 0.92.

In comparing these two pairs of estimates it is important to appreciate that the two sets of data were for the first and the second set of 40 cases from the same domain, generated independently under a probability model that implied the *correct probabilities* for those two cases to be 0.25 and 0.83, respectively. The third set gave 0.20 and 0.19 (*sic*), while the fourth one gave 0.44 and 0.83; etc. – all of these results derived under the same model that implied the correct probabilities.

The lesson from this is very discouraging: the statistical estimates of the diagnostic probabilities, while objective, are very unstable – they have *very low reproducibility* – in repeat experiences; that is, for reliable estimates needed is very large experience. From the first 400 cases, the estimates for those two new cases became 0.26 and 0.81 respectively. (Cf. the correct values above.) And to be noted here is that the larger is the number of parameters in the model fitted to the data, the larger needs to be number of cases (data points) for any given level of reproducibility/precision of the result. In the example here, the model involved only five parameters, but the results from sets of 40 cases were very imprecise nevertheless.

The statistical model underlying, and fitted to, the data was quite typical for a situation like this: logistic for the probability that I is present, with the linear compound for the logit of this probability (of $Y = 1$) formulated as $B_0 + B_1X_1 + B_2X_2 + B_3X_3 + B_4X_2X_3$, where X_1 and X_2 are indicators of D_1+ and D_2+, respectively, and $X_3 = D_3 - 100$.

The data on D_1, D_2, and D_3 were generated under a stochastic model for their joint distribution. Pivotal in this was the distribution of X_2 (the indicator of D_2+) as Bernoulli (0.5). X_3 was Gaussian with mean $100 + 20X_2$, variance 100; and X_1 was equal to X_2 with probability to 0.75 (otherwise $X_1 = 1 - X_2$).

For the generation of data on the presence/absence of I, the parameters B_0 through B_4 in the logistic model for $\Pr(Y = 1)$ were –1.1, 0.0, 1.1, 0.05, and 0.03, respectively (with the Xs as defined above).

. . .

As at issue here is diagnostic *research*, some notes on the *validity* of this type of experience as a source of evidence about the magnitudes of the parameters (*B*s) in the statistical model are in order. Apart from the correctness of the data as such, the question here is about the validity of the set of cases of record; and in this, there are two issues: the validity of the subset of the known cases from the domain such that the truth about the I status got to be known (to some doctors), and the validity of the subset of these informative cases that got to be part of the experience – first- or second-hand – being used for learning by a particular doctor.

The learning set has not lost validity on account of the truth having become known with probability that depends on the diagnostic indicators that are being accounted for. But the set of cases that constitutes the basis of learning obviously must not be select on the basis of the known I status itself, nor on the basis of correlates of this not accounted for in the model.

There are subtleties regarding the validity of the database for diagnostic research beyond the outlines of this above, but these are not a concern in this introduction to medicine rather than to medical research. However, the student and the practitioner concerned to learn about diagnostic probabilities through experience in practice, informally, without statistics, needs to understand that the issues of validity of the learning set of cases are involved in this informal framework as well.

Upon this brief visit to the domain of diagnostic research against the backdrop of what came up in Section 8.1 in regard to the 'tacit knowledge' about diagnostic probabilities gleaned from diagnostic experience, it should be clear that learning diagnosis – diagnostic probability-setting – via the intermediary of diagnostic research is possible, while learning it informally from practice experience is quite illusory. Thus, while patients need to seek doctors' learned help to understand the implications of such sickness as they are experiencing, doctors need diagnostic researchers' help to understand the implications of the cases they've encountered or are otherwise familiar with. But ideally they would have the requisite knowledge-base of diagnoses from the diagnostic species of quintessentially applied medical science (Section 1.4 and Appendix 3).

Appendix 3: Neo-Hippocratic Learning of Medicine

In ancient Greece, the art of Hippocratic medicine was taught – solo – by masters of it. And their students, in turn, felt an obligation to propagate the learning by teaching the art to students of their own (Section 6.4).

In a modern country, no one can be a master of 'the art of medicine,' qualified to teach all of modern medicine. 'The art of medicine' has become a hollow concept: that art no longer exists; it has evolved into various component arts/disciplines of medicine. (Cf. Preface and Section 1.5.)

The modern counterpart of the Hippocratic teaching of medicine in ancient Greece would be the teaching of a *particular, suitably-defined discipline* of medicine (Section 7.2) by a single master of this art or by a set of these (interchangeable) masters. Thus, a *singular type of teacher* would again be the source of all of the learning, though now specific to whichever particular discipline of modern medicine, following whatever propaedeutic studia generalia of medicine (Preface) and study of the concepts and principles specific to the discipline in question.

In this modern counterpart of the Hippocratic teaching of medicine there would be one *major novelty beyond the focus and content* of it. The Hippocratic art was originally propagated by oral teachings alone (the 'Hippocratic' corpus of writings developing only later). By contrast, the relevant educational content of (i.e., the knowledge-base of practice in) a modern discipline of medicine – focusing on, say, diagnoses in a particular segment of emergency medicine (cf. Section 7.2) – could be *codified in its entirety* and even made *accessible as needed* in the course of practice (through an 'expert system' imbedded in cyberspace; Section 5.6).

Hippocrates – this outstandingly great teacher – was a practicing doctor; and in the neo-Hippocratic framework of learning medicine, similarly, the discipline-specific teacher would be a practicing doctor in the discipline in question and, as such, a modern 'role model' for the student about the practice proper – about the deployment of the discipline's 'expert system' etc.

The teaching-and-learning following all of the relevant preliminaries ideally would take place, ultimately, in the teacher's actual practice, even though it initially would be conducted in the medical equivalent of the flight-simulator setting of airline-pilots-in-training: As of the moment a case from the discipline's domain – possibly one with a particular species of gnosis as its sole concern (Section 7.2) – would come up, it would be classified according to the categories of presentation in terms of which the system is organized (e.g., a diagnostic system according to species of sickness; Section 5.2). Upon entry of the presentation category, the system would assume its role as 'the expert.'

The system would ask questions and the doctor would answer these, based on facts ascertained on the case. And based on these inputs together with the general medical knowledge codified in the system – in terms of gnostic probability functions (cf. Appendix 1) – 'the expert' would give the relevant gnostic probabilities to the doctor – to form the basis of the doctor's teaching the client about their health, in the prognostic context as an input to their decision about the choice of treatment (Section 5.7). Upon these gnoses, in the context of actual practice in particular, the master would teach the client on the basis of the gnoses – and teach the student about such teaching, on a case-by-case basis.

Given that the requisite knowledge for gnoses would be codified in the expert system, the student would not need to personally acquire and memorize that knowledge (which would be impossible). In consequence of this, as noted above, the discipline-specific learning would focus on substantive concepts and principles specific to the discipline at issue – perhaps also imbedded in the discipline's expert system (along with the facts-acquisition algorithms and the facts-conditional gnoses) – together with the *skills* needed in the ascertainment of the gnosis-relevant clinical (rather than laboratory-based) facts on the cases that may come up.

Regrettably, in my view, this has not been the result of the evolution of the teaching and learning of medicine from the time of Hippocrates to the present in this Information Age. In fact, the evolution has taken away from the focus on and development of the knowledge-base (gnostic) of medicine, as the teachers, and to an extent, through their influence, the students too, are preoccupied with and distracted by 'basic' medical research and its results in addition to and even instead of medicine itself (Section 2.4), while quintessentially applied, gnosis-oriented medical research has not been seriously cultivated (Section 1.4.2).

And so in closing here I leave to modern students of medicine one final proposition to weigh and consider (à la Francis Bacon; Section 2.2). It is predicated on the premise that quest for personal happiness animates all actions of humans (Sections 2.3 and 5.8.2), and that happiness in professional life flows from the virtue (Aristotelian) of excellence in it (Sections 6.1 and 6.3). My final proposition here is this:

A student aiming to achieve *excellence* in medicine could more realistically expect to find it – and its consequent *professional happiness* – in the direction of the neo-Hippocratic learning and practicing I sketched above as a dream about the future, than in the framework of the prevailing realities of medical education (Section 2.4) and medicine proper. For, in medicine as it still is, a major detraction from professional happiness continues to be the proto-Hippocratic frustration with the art (per the aphorism in Section 1.1) – how long the study of it is as a proportion of the length of life post entry into the study; how treacherous experience in practice is as a teacher of the knowledge-base of the art (Section 8.1); and how difficult in the practice of the art is the use of judgements as a substitute for knowledge about the relevant truths (Sections 5.2, 5.3, and 5.4). But in the future I'm dreaming of (for time beyond mine), study of one's art of medicine would not be unduly long (due to focus on the relevant, as sketched above); treacherous experience of practice would not have to be relied on as the source of 'tacit knowledge' (cf. above); and ad-hoc judgements would be replaced by objective knowledge (accessed from 'expert systems'; cf. above) – so that excellence of the deployed knowledge-base would be the cornerstone of overall excellence in every doctor's professionalism (cf. Sections 9.1 and 9.2).

Appendix 4: Answers to the Questions in the Preface

Q 1: Are 'disease,' 'sickness,' and 'illness' synonyms?

A 1: These three terms are commonly used as though they were synonyms, but in critical usage they are not. 'Illness' is the antonym of 'health.' Sickness is what a person with a case of a hidden illness-definitional somatic anomaly may directly suffer from (or the unwellness may have an extrinsic direct cause). Disease, like defect and trauma, is a particular species of illness; it (different from defect) is a process-type somatic anomaly, and its pathogenesis (different from that of trauma) is intrinsic to (patho)biology. (Sections 3.2, 3.3, and 3.4).

Q 2: Are 'treatment,' 'therapy,' and 'intervention' synonyms?

A 2: These three terms, like those in Q 1, are commonly used as though they were synonyms, but in critical usage their denotations are distinct: Treatment (in medicine) is therapy only when intended to change the course of an existing case of illness (and not merely sickness from it) for the better. And treatment is intervention (therapeutic or preventive) only when it is an action directly on the client's soma (rather than a change in their environment or behavior). (Section 4.6).

Q 3: What is true and unique about all genuine disciplines of medicine (distinguishing them from all paramedical disciplines, i.a.)?

A 3: True and unique about – and hence definitional to – all genuine disciplines of medicine is the doctor's pursuit and (potential) attainment of first-hand esoteric knowing – gnosis (dia-, etio-, and/or prognosis) – about the health of the client, as the basis of all else in his/her services to the client (most notably teaching the client about their health beyond the available facts pertaining to this, by bringing general medical knowledge to bear on those facts). (Section 4.1.1).

Q 4: Can a modality of treatment – surgery, say – be definitional to a genuine discipline of medicine?

A 4: Not justifiably. Disciplines of medicine are rationally defined according to what the gnoses in them are about; and under prognosis, all possible modalities are considered (Section 4.1.2). A discipline said to be one of 'surgery' is a discipline of medicine insofar as the relevant gnoses are in the essence of it and the use of surgical intervention by the doctor only an incidental element in it. (Sections 4.1.1, 4.1.2 and 9.1).

Q 5: What logically is the essence of diagnosis, and what is the source of its requisite knowledge-base?

A 5: Contrary to what is said in dictionaries of medicine, diagnosis cannot logically be said to be the determination of what illness is at the root of the patient's sickness. For one, it is not a generally-tenable premise that the client's complaint about sickness actually is a factual one and, as such, a manifestation of illness (instead of having an extrinsic direct cause). And even when it is, the set of available facts constituting the diagnostic profile as the ad-hoc input to diagnosis is not, generally, a sufficient basis for such a determination (i.e., pathognomonic about the nature of the illness underlying and causing the sickness). Logically, thus, the essence of diagnosis is knowing about the correct probability that a case of a particular hidden illness is present, given the diagnostic profile of the case. (Section 4.2.1).

The requisite knowledge-base of diagnosis consequently is, ultimately, about the probabilities of the presence of the various possible underlying illnesses, conditional on the diagnostic profile (in the form of diagnostic probability functions; Appendix 1). The source of 'tacit knowledge' about these probabilities is commonly thought to be the doctor's experience with cases presenting for diagnosis in his/her diagnostic practice. Belief in such practice-based learning about diagnostic probabilities is, however, quite unrealistic, the source of true diagnostic knowledge being diagnostic research. (Sections 1.4, 8.1 and 8.2).

Q 6: What are tenable conceptions of scientific medicine and medical science, respectively?

A 6: While there now are two eminent conceptions of scientific medicine (Section 4.1.5) and many others besides (Section 6.5), the only tenable conception of it is: medicine with a logical theoretical framework and, in such a framework, knowledge-base from (medical) science (Section 4.1.5).

Medical science is science aimed at advancement of medicine (its practice; Section 1.4). The intended advancement is either the availability of new 'tools' useful for medicine or improvement of the knowledge-base of medicine (Section 1.4).

Q 7: What is the essence of ethical medicine?

A 7: Medicine is ethical insofar as the doctor is worthy of the client's trust that the service provided by the doctor is, in every respect, excellent – that is, as good as anyone could provide. The essence of ethical medicine is the doctor's professional trustworthiness in this sense (Sections 6.3 and 7.3).

Q 8: How is professional happiness in medicine best assured?

A 8: This question is specifically about a physician's professional happiness as a doctor, and this needs to be thought of as his/her happiness about the quality – excellence (cf. A 7 above) – of his/her professional services to the clients.

Toward this happiness-producing virtue (Aristotelian; Section 6.1) needed first is commitment of excellence as a doctor, the first-order feature of this being uncompromised dedication to serving no ends other than the best interests of the clients (Sections 6.3 and 7.1). Given this commitment, the path to that goal of excellence begins with suitable definition of one's domain of the pursuit of excellence (Sections 1.5 and 7.2). And then, in the pursuit proper, the key is wisdom in the meaning of realism (Sections 1.1, 8.1, and 8.2). (In the future, I hope, all of this will be very different; Appendix 3 above).

Printed in the United States
By Bookmasters